# Policy Legitimacy, Science and Political Authority

Voters expect their elected representatives to pursue good policy and presume this will be securely founded on the best available knowledge. Yet when representatives emphasize their reliance on expert knowledge, they seem to defer to people whose authority derives, not politically from the sovereign people, but from the presumed objective status of their disciplinary bases.

This book examines the tensions between political authority and expert authority in the formation of public policy in liberal democracies. It aims to illustrate and better understand the nature of these tensions rather than to argue specific ways of resolving them. The various chapters explore the complexity of interaction between the two forms of authority in different policy domains in order to identify both common elements and differences. The policy domains covered include: climate geoengineering discourses; environmental health; biotechnology; nuclear power; whaling; economic management; and the use of force.

This volume will appeal to researchers and to convenors of post-graduate courses in the fields of policy studies, foreign policy decision-making, political science, environmental studies, democratic system studies, and science policy studies.

**Michael Heazle** is an Associate Professor with the Griffith Asia Institute and the Griffith University School of Government and International Relations, Australia. His teaching and research interests include international relations, politics, and the treatment of uncertainty in foreign and domestic policy making.

**John Kane** is a Professor with the Centre of Governance and Public Policy and the School of Government and International Relations, Griffith University, Australia, where he researches and teaches in the fields of political theory, political leadership and US foreign policy.

**Science in Society Series**

Series Editor: Steve Rayner

*Institute for Science, Innovation and Society, University of Oxford*



The Earthscan Science in Society Series aims to publish new high-quality research, teaching, practical and policy-related books on topics that address the complex and vitally important interface between science and society.

**Vaccine Anxieties**
Global Science, child health and society
*Melissa Leach and James Fairhead*

**Democratizing Technology**
Risk, responsibility and the Regulation of Chemicals
*Anne Chapman*

**Genomics and Society**
Legal, ethical and social dimensions
*Edited by George Gaskell and Martin W. Bauer*

**A Web of Prevention**
Biological weapons, life sciences and the governance of research
*Edited by Brian Rappert and Caitrìona McLeish*

**Nanotechnology**
Risk, ethics and law
*Edited by Geoffrey Hunt and Michael Mehta*

**Unnatural Selection**
The challenges of engineering tomorrow's people
*Edited by Peter Healey and Steve Rayner*

**Debating Climate Change**
Pathways through argument to agreement
*Elizabeth L. Malone*

**Business Planning for Turbulent Times**
New methods for applying scenarios
*Edited by Rafael Ramírez, John W. Selsky and Kees van der Heijden*

**Influenza and Public Health**
Learning from past pandemics
*Tamara Giles-Vernick, Susan Craddock and Jennifer Gunn*

**Animals as Biotechnology**
Ethics, sustainability and critical animal studies
*Richard Twine*

**Uncertainty in Policy Making**
Values and evidence in complex decisions
*Michael Heazle*

**The Limits to Scarcity**
Contesting the politics of allocation
*Lyla Mehta*

**Rationality and Ritual**
Participation and exclusion in nuclear decision making, 2nd Ed.
*Brian Wynne*

**Integrating Science and Policy**
Vulnerability and resilience in global environmental change
*Edited by Roger E. Kasperson and Mimi Berberian*

**Dynamics of Disaster**
Lessons on risk response and recovery
*Rachel A. Dowty Beech and Barbara Allen*

**The Social Dynamics of Carbon Capture and Storage**
Understanding CCS Representations, Governance and Innovation
*Edited by Nils Markusson, Simon Shackley and Benjamin Evar*

**Science and Public Reason**
*Sheila Jasanoff*

**Marginalized Reproduction**
Ethnicity, infertility and reproductive technologies
*Edited by Lorraine Culley, Nicky Hudson and Floor van Rooij*

**Resolving Messy Policy Problems**
Handling conflict in environmental, transport, health and ageing policy
*Steven Ney*

**The Hartwell Approach to Climate Policy**
*Edited by Steve Rayner and Mark Caine*

**Reconstructing Sustainability Science**
Knowledge and action for a sustainable future
Thaddeus R. Miller

**Experiment Earth**
Responsible innovation in geoengineering
*Jack Stilgoe*

**Rethinking Greenland and the Arctic in the Era of Climate Change**
New Northern Horizons
*Frank Sejersen*

**International Science and Technology Education**
Exploring Culture, Economy and Social Perceptions
*Edited by Ortwin Renn, Nicole C. Karafyllis, Andreas Hohlt and Dorothea Taube*

**Policy Legitimacy, Science and Political Authority**
Knowledge and action in liberal democracies
*Edited by Michael Heazle and John Kane*

**Systems Thinking for Geoengineering Policy**
How to reduce the threat of dangerous climate change by embracing uncertainty and failure
*Robert Chris*

'As the academic world lurches from its love affair with postmodernism to whatever comes next, the relationship between politics and science is going to be crucial. This book is a welcome contribution to the discussion with the huge and refreshing advantage that not everything is cast as a choice between technocracy and democracy.'

*Harry Collins FBA, Distinguished Research Professor, Cardiff University, UK*

'This book's engaging case studies show how the role of experts in politics differs dramatically in different policy areas. Rather than merely debunking or defending expertise, the authors examine the subtle processes through which scientific and political authority shape each other and yet remain distinct. In a time of widespread concern over the politicization of science, Heazle and Kane offer sensible and much needed guidance.'

*Mark B. Brown, Professor, Department of Government, California State University, USA*

# Policy Legitimacy, Science and Political Authority

## Knowledge and action in liberal democracies

Edited by
Michael Heazle
and John Kane

Routledge
Taylor & Francis Group
LONDON AND NEW YORK

earthscan
from Routledge

First published 2016
by Routledge
2 Park Square, Milton Park, Abingdon, Oxon OX14 4RN

and by Routledge
711 Third Avenue, New York, NY 10017

*Routledge is an imprint of the Taylor & Francis Group, an informa business*



*British Library Cataloguing-in-Publication Data*
A catalogue record for this book is available from the British Library

*Library of Congress Cataloging-in-Publication Data*
Policy legitimacy, science and political authority: knowledge and action in liberal democracies/edited by Michael Heazle and John Kane.
pages cm. – (Earthscan science in society) 1. Policy sciences.
2. Expertise – Political aspects. 3. Science – Political aspects.
4. Democracy. I. Heazle, Michael. II. Kane, John, 1945 April 18–
H97.P6457 2016
320.6–dc23 2015016573

ISBN: 978-1-138-91907-5 (hbk)
ISBN: 978-1-315-68806-0 (ebk)

Typeset in Goudy
by Florence Production, Stoodleigh, Devon, UK

# Contents

# Preface and acknowledgements

This volume is the product of a 2013 workshop entitled 'Good Public Policy: On the Interaction of Political and Expert Authority', convened by the editors on behalf of the Griffith University Centre for Governance and Public Policy and hosted in Washington, DC by Arizona State University's Consortium for Science, Policy & Outcomes (CSPO) offices.

The editors would like sincerely to thank all the contributors for their chapters, numerous revisions, and time spent helping craft the collection into a cohesive whole. We would particularly like to thank Haig Patapan for his initial advice on the project and contribution to the Introduction, and also Dan Sarewitz, Steve Rayner, and Darrin Durant for their invaluable conceptual advice and feedback on various drafts of the Introduction and Conclusion; any remaining errors or omissions are ours alone. We also acknowledge Dan's terrific support in hosting the workshop at CSPO.

We must also thank Daniela di Piramo for her excellent work in preparing the manuscript, chasing down sources, and looking after the myriad little jobs that constantly pop up in the course of preparing edited collections for publication.

Finally, we would like to acknowledge the support of the Australian Research Council whose funding made this project possible.

Michael Heazle and John Kane
Brisbane, April 13, 2015

# Contributors

**Darrin Durant** lectures on science and policy making at the University of Melbourne. He has published widely on the general problem of incorporating both experts and publics in deliberative exercises and decision-making forums. He has also published extensively on technical and socio-political debates about nuclear waste disposal. Normatively speaking, other work on energy policy trajectories has voiced skepticism about the wisdom of nuclear power. Recent work has taken up the climate change issue, focusing on Australia's somewhat unique situation of having a prominent climate change organization that was once a formal part of government but was sacked and is now out there in the wild trying to save the planet.

**Michael Heazle** is an Associate Professor with the Griffith Asia Institute and the Griffith University School of Government and International Relations where he teaches International Relations and Politics. From 1992 to 2000, Dr Heazle was a regular contributor to the *Far Eastern Economic Review*, and wrote for a number of other domestic and international media. Dr Heazle has researched and published in the areas of energy, human, and environmental security; policy making and the treatment of specialist advice; and China–Japan relations. His works include a collection of books and edited volumes with various university presses and publishers (University of Washington Press, Cambridge University Press, Earthscan, Edward Elgar) and research articles in peer reviewed journals including *Marine Policy*, *Environmental Science and Policy*, *Intelligence and National Security*, and the *Australian Journal of International Affairs*.

**Clare Heyward** is a Leverhulme Early Career Fellow at the University of Warwick. Her current project is *Global Justice and Geoengineering*. Previously, she was a James Martin Research Fellow on the Oxford Geoengineering Programme. Clare is interested in issues of global distributive justice and intergenerational justice, especially those connected to climate change.

**John Kane** (Ph.D., LSE) is Professor in the School of Government and International Relations, Griffith University, Australia, and in Griffith's Centre for Governance and Public Policy. He teaches in political theory, leadership and US foreign policy and is the author of *The Politics of Moral Capital* (Cambridge University Press, 2001) and *Between Virtue and Power: The Persistent Moral Dilemma of US Foreign Policy* (Yale University Press, 2008). He is also co-author (with H. Patapan) of *The Democratic Leader: How Democracy Defines, Empowers and Limits its Leaders* (Oxford University Press, 2012) and co-editor of *Good Democratic Leadership: On Prudence and Judgment in Modern Democracies* (Oxford University Press, 2014).

**David Kriebel** received his master's degree in physiology/occupational health (1983) and doctorate in epidemiology (1986) from the Harvard School of Public Health. He did post-doctoral work on exposure assessment for epidemiology with Dr Tom Smith at the University of Massachusetts Medical Center, and spent a year as a scholar in residence at the Center for the Study and Prevention of Cancer in Florence, Italy on a Fulbright Fellowship. Since 1988, he has been on the faculty of the Department of Work Environment, University of Massachusetts Lowell, where he holds the rank of full Professor. Dr Kriebel is also the Director of the Lowell Center for Sustainable Production, which collaborates with industries, government agencies, unions, and community organizations on the redesign of systems of production to make them healthier and more environmentally sound. Dr Kriebel's research focuses on the epidemiology of occupational injuries, cancer, and non-malignant respiratory disease. He has co-authored two textbooks: *Research Methods in Occupational Epidemiology* with Harvey Checkoway and Neil Pearce (Oxford University Press, 2004) and *A Biologic Approach to Environmental Assessment and Epidemiology*, with Thomas J. Smith (Oxford University Press, 2010).

**Haig Patapan** is Director of the Centre for Governance and Public Policy and Professor in the School of Government and International Relations, Griffith University, Australia. His research interests are in democratic theory and practice, political philosophy, political leadership and comparative constitutionalism.

**Paul R. Pillar** is Non-resident Senior Fellow at the Center for Security Studies of Georgetown University. He is also a Non-resident Senior Fellow at the Brookings Institution and an Associate Fellow of the Geneva Center for Security Policy. He retired in 2005 from a 28-year career in the US intelligence community. His senior positions included National Intelligence Officer for the Near East and South Asia, Deputy Chief of the DCI Counterterrorist Center, and Executive Assistant to the Director of Central Intelligence. He is a Vietnam War veteran and a retired officer in the US Army Reserve. Dr. Pillar received an A.B. *summa cum laude* from Dartmouth College, a B.Phil. from Oxford University, and an M.A. and Ph.D. from Princeton University. His books include *Negotiating Peace: War Termination as a Bargaining Process* (1983), *Terrorism and U.S. Foreign Policy* (2001), and *Intelligence and U.S. Foreign Policy: Iraq, 9/11, and Misguided Reform* (2011). He blogs at nationalinterest.org.

**Steve Rayner** is James Martin Professor of Science and Civilization and Director of the Institute for Science, Innovation and Society at Oxford University, where he also co-directs the Oxford Geoengineering Programme. He previously held senior research positions in two US National Laboratories and has taught at leading US universities. He has served on various US, UK, and international bodies addressing science, technology and the environment, including Britain's Royal Commission on Environmental Pollution, the Intergovernmental Panel on Climate Change and the Royal Society's Working Group on Climate Geoengineering. In 2008 *Wired* magazine included him on its Smart List as 'one of the 15 people the next US President should listen to'.

**Daniel Sarewitz** is Professor of Science and Society, and co-director and co-founder of the Consortium for Science, Policy, and Outcomes (CSPO), at Arizona State University (www.cspo.org). His work focuses on revealing and improving the connections between science policy decisions, scientific research and social outcomes. His most recent book is *The Techno-Human Condition* (co-authored with Braden Allenby; MIT Press, 2011). He is editor of the magazine *Issues in Science and Technology* (www.issues.org) and is also a regular columnist on science policy affairs for the journal *Nature*.

**Paul J. Scalise** (Ph.D., Oxford University: 2009) is Senior Research Fellow at the IN-EAST Institute of Advanced Studies, University of Duisburg-Essen, Germany, and Research Manager of its 'Institutional Innovations in East Asian Energy and Low-Carbon Markets' group. A former JSPS Research Fellow at The Institute of Social Science, University of Tokyo, Dr Scalise is also an Associate Fellow at the Institute for the Analysis of Global Security and Non-Resident Fellow at the Institute of Contemporary Asian Studies, Temple University, Japan Campus.

# Figures and tables

## Figures



## Table

# 1 Good public policy

## On the interaction of political and expert authority

*Michael Heazle, John Kane and Haig Patapan*

This book examines the tension between political and expert authority in order to consider its effects on the design and implementation of good public policy in a modern, liberal democracy. What counts as good policy in specific instances is often contentious, but voters presume, at the very least, that it will be policy securely founded on the best available knowledge. Yet when our elected political representatives, in the exercise of their mandated authority, emphasize their reliance on expert knowledge, they seem to defer to people whose authority derives, not politically from the sovereign people, but from the presumed objective status of their disciplinary bases. This creates the potential for recurring tension between the two interacting sources of authority.

The resolution of such tension may be more or less difficult. In some cases tension may hardly be evident, being resolved by astute exercises of political judgment. If, however, a contest arises over the merits of a particular policy, attention will almost invariably focus on the nature of the expert authority that sustains it. Experts are then exposed to political scrutiny, even attack, as debates over good policy turn into contests over scientific or professional judgment. In such cases, expert and political authorities become so thoroughly entangled that it becomes difficult to separate them.

To understand the elements of good public policy making in a democracy, therefore, it is necessary to examine the interaction of political and expert authorities in order to see where they may be in accord or, more significantly, when and why they may clash, and what the outcomes might be in either case. This book explores this issue through eight case studies drawn from a broad spectrum of significant and complex policy domains. Taken together, these studies provide a subtle and detailed examination of the possible tensions that may exist between expert and political forms of authority, allowing us to better understand how they may be resolved, or why they may remain unresolved, in modern democracies.

We begin by making a general point on terminology. We will be speaking most often of 'expert authority' but occasionally of 'scientific authority' and sometimes of the issue of 'politics versus science'. Governments, of course, draw on a vast range of expertise from the hard sciences, to engineering science, through biological and medical sciences to the social sciences not to mention legal science (as it is sometimes called) as well as on intelligence analysts, accountants and many other varieties of professionals. The practitioners of these disciplines can hardly be assimilated one to another, needless to say, but may be sufficiently grouped together as 'experts' by being, in the modern post-Weberian age, informed by common values of rational inquiry that emphasize methodologies adapted to subject matter, the careful acquisition and compiling of evidence, its critical analysis in forming any conclusions, and so on. What is held to distinguish them as a group from political actors is that the legitimacy of experts derives explicitly from their adherence to these rational values, and not from the political values embedded in any policy field they may seek to inform. Indeed, attempts by experts to 'overextend' their authority into the realm of making, as opposed to informing, policy priorities and choices clouds this important distinction and ultimately undermines the authority of their expertise. The validity of this distinction has, as we will see, become subject to critique over recent times, but it remains conceptually clear enough for us to be able to speak broadly of 'experts' whose authority can support but may also often be in tension with political authority.

## The significance of regime

One way of trying to understand the tension between political and expert authority is by considering the nature of political regimes. We may conceive, for example, of purely technocratic and purely ideological regimes as ideal types situated at either end of a continuum of combinatorial possibilities with respect to appeals either to expertise or ideology in grounding legitimacy (Hoppe 2005: 206–08). Singapore, for instance, as a one-party state where expert-based solutions are valued over pluralism and political debate in determining policy ends and means, would be situated toward the technocratic end of the scale. Maoist China (1949–1976), in contrast, would appear very close to the opposite end as policies during this period (such as the Great Leap Forward and Cultural Revolution) were almost entirely a function of the Communist Party's ordering of ideological values.

There are, however, clear limitations in this way of understanding the tension between political and expert authorities. It is hardly imaginable that a technocracy can do away with politics altogether, though it may successfully occlude them from the public gaze. It is true that expert authority is founded on the possession of knowledge held to be technical, rational, objective and thus essentially apolitical, which is why technocratic government is usually conceived of as 'apolitical' government. Yet it would be more accurate to say that expert authority is taken by technocrats as the central ground of their *political* authority, effectively reducing the politics of choice to expert judgments about 'what works'. Such a ground of political authority can be distinguished from those assumed by other, very different regimes—for example, bloodline descent in monarchies, a mandate of heaven in imperial China, a strongman's capacity to provide order in fascism, a vanguard's leadership of a progressive movement in Leninism, and electoral success in democracies.

If, on the other hand, we consider the other end of our proposed spectrum, it is not clear that any regime can entirely dispense with technological expertise. It may be possible to imagine one so grounded in ideology that rationalistic authority plays no role whatsoever, political choices about both ends and means being made not on the basis of what pragmatically works but purely on value premises of 'what *should* be'. But such ideological blinkeredness seems inevitably to lead to economic or political disorder and a consequent turn toward more pragmatic solutions, making the regime increasingly reliant on technical-expert rather than ideological authority.[1]

There is, therefore, an important question about the extent to which any regime must inevitably confront political-expert tension, however it may seek to conceal or disguise the matter. Nevertheless, the regime approach is useful in revealing an important insight into liberal democracies, which presumably sit somewhere in the middle of our notional spectrum. Liberal democracies seem to demand the 'best of both worlds' from their governments—an assured mandate to govern *plus* rationally defensible policies.

Democratic leaders, of course, rely fundamentally on electoral legitimacy for their political authority, the condition of which legitimacy is that they are never the people's rulers but notionally always their 'representatives', governing with the people's consent. This authority is constrained on one hand by law (the liberal element of a liberal democracy) and on the other by the permanent democratic requirement of continuous public accountability. Democratic legitimacy is, in fact, a rather fragile commodity that must be constantly shored up and renegotiated by elected leaders as they attempt to govern while casting their gaze toward the next election (Kane and Patapan 2012). An electoral mandate is merely the bedrock of democratic legitimacy upon which a structure of performance-based legitimacy must be built. Consent is granted for the sake, not just of representative government, but of *good* government, meaning that policy must be justified as best promoting the public and national interest.

However, good policy making implies a capacity for astute judgment as well as a level of expertise in relevant fields, which democratic representatives do not necessarily (or even commonly) possess. It should be remembered that no formal qualifications are required to become a democratic representative, merely the capacity to be elected. But even able and educated representatives will be deeply reliant on the services of various kinds of experts from within and outside government—economists, agronomists, lawyers, medical specialists, educationalists, engineers, meteorologists, marine biologists, environmentalists, social scientists, and so on—who, whether employed directly by government or not, typically provide both policy advice and regulatory oversight. So while political authority in a liberal democracy remains essentially democratic, governments in a scientific age must also lean heavily on a class whose authority comes, not from politics, but from Enlightenment-based claims to the possession of various forms of specialist knowledge.

The policies of leaders will usually appear fully legitimate, then, only if bolstered by the authority of relevant experts of many kinds across multiple issue areas. Expert authority, in other words, is expected to provide for the rationalization of policy and the depersonalization of political power (Ezrahi 1990; Hisschemoller 2005: 93). Liberal democracies thus, by their very nature, reveal most fully and comprehensively the tension between political and expert authority that is often concealed or obscured in other regimes. At issue in this volume is the exact nature of this tension and its general implications for good public policy.[2] We turn now to recent scholarship that has been centrally concerned with such questions.

## Three waves of thinking on political-expert tensions

Over the last six decades scholars have produced a rich and growing body of conceptual and empirical research seeking to clarify the nature of expert knowledge and explain its role in liberal democratic policy making. Collins and Evans (2002) framed the evolution of ideas over this period as a series of three waves, which, though criticized as too reductionist a representation (see, for example, Jasanoff 2003), nevertheless provides a succinct overview of changing assumptions about the nature and interaction of expert and political forms of authority across time.

The 'waves' are useful from our perspective because they help us identify a core limitation of the general scholarship—namely, its normative desire to resolve the tension between political and expert authority in favor of democracy, which tends (we claim) to obscure the complex reality of their interrelationship across different domains. First wave scholarship endorses the common view of politics and science as distinct, uncritically accepting the ability of experts to produce objective truths about the world, and presumes no tensions exist so long as policy makers straightforwardly apply consensual scientific knowledge. The second wave interpreted wave one, however, as exhibiting excessive deference to experts who were capable of framing debates in a top-down manner by their *ex cathedra* judgments, and argued the need to open up policy to broader involvement. Wave two scholars claimed, indeed, that the distinction between expert and political authority was largely spurious, that in fact all authority is inherently 'constructed' and therefore fundamentally political. Third-wave scholars took the tension between expert and political authority more seriously, but agreed with wave two that broader involvement in policy making, through recourse to certain democratic initiatives such as deliberative and consultative practices, is needed to ensure democratic legitimacy.

Collins and Evans, in their 2003 reply to critics (Collins and Evans 2003), described the third wave as a form of 'policy turn' in which attention shifts from case study deconstructions of science/expertise to a broader focus on better understanding the intersection of expertise and policy making with a view to good policy making. Some scholars in this turn took the expertise-democratic politics tension quite seriously, but disagreed about the nature and status of expertise and the sufficiency of deliberative initiatives to resolve the tension. Others, however, remained inclined to dissolve expertise into politics and wrote as if politicized decisions were enough. Others again stopped short of this but nevertheless sought to disempower experts almost completely in the policy process. Collins and Evans, in contrast, conceding that their own work was a contribution to this policy turn, explicitly maintained that expert authority could not be reduced to a type of political authority, that deliberative mechanisms were insufficient, and that the answer to managing the tension lies in finding a rationale for valuing science even though we know it is not perfect.

First-wave thinking, which dates from the early post-war period, sought to further entrench Enlightenment ideas of objective knowledge and its contribution to rational policy making (see, for example, Lane 1966). According to this 'classical-modernist' view (Hajer 2003), the roles of experts and government are clearly prescribed by the principles and institutions of liberal democracy, with experts providing the best available technical solutions and advice to governments as the latter rationally respond to the policy needs of the day. According to this division of labor, government must refrain from intruding on expert autonomy and experts must refrain from advocating policy. If tensions do arise, it is a result of one or other of the players, policy makers or experts, stepping beyond their properly defined roles. But in the event of a conflict arising between the two, political authority is expected, as Churchill once put it, to remain 'on top' while experts stay 'on tap'.[3] In practice there will (or should) be a cooperative merging of political and expert authorities to produce rational, 'evidence-based' policy solutions to problems framed by both expert knowledge *and* liberal ideas and values. Even while cooperating, however, each of these authorities remains autonomous of, and thus potentially in competition with, the other because they are conceptually and constitutionally distinct.

Even within this first-wave tradition of policy studies, there was some critical analysis of the linear-rationalist policy models that the Enlightenment view underpinned (Lindblom 1959, 1979; March and Olsen 1989; Fischer and Forester 1993). From the late 1960s, however, an emerging social constructivist perspective on science and the kind of knowledge it produced presented a deeper challenge to what was seen as the old positivist view of expert authority. This form of critique was built in part on earlier efforts within the philosophy of science to explain the rationality of scientific knowledge, but which paradoxically threw doubt on science's claims to objectivity and political neutrality (Kuhn 1996; Popper 1972; Lakatos in Larvor 1998; Feyerabend 1976). The attraction of such skeptical perspectives was socially enhanced by the declining public prestige of experts after the 1960s, as the disbenefits of technology became more apparent (Beck 1992; House of Lords 2000: Chapters 1–2; Collins and Evans 2002; McClay 2009; Hughes 2012). Expert authority suffered because some applied technologies were perceived as dangerous (nuclear technology), unreliable (climate science), undesirable (genetically modified food), or morally questionable (stem cell research). Also encouraging skepticism was the increasingly pervasive democratic ethos of contemporary society, with its rise of 'consumer citizens' and massively greater public availability of information and analysis.[4] McClay (2009: 145) explained the change by pointing to the essentially undemocratic character of expert authority:

> The authority of experts took a beating in the past 40 years precisely because the idea toward which such authority tended—the notion that expert knowledge could be amassed and exercised in an entirely disinterested way—was deeply mistaken, and deeply subversive of any meaningful conception of democracy.

The second wave of social constructivist critique dismissed the first wave's positivist claims to privileged scientific knowledge and developed many of the core assumptions and methods of contemporary Science and Technology Studies (STS). Major issues included: how, if at all, genuine 'expert' knowledge can be distinguished from other types of knowledge claims for the purposes of policy (Collins and Evans 2002; Kitcher 2011); why demarcation between scientific and non-scientific forms of knowledge is less about claims to unique practice than about ideology and rhetoric (Gieryn 1983; Wynne 1996); how social and political influences cause knowledge and evidence to be co-produced (Jasanoff 1990, 2004); how technical expertise and technology can be invoked to limit political choice and obscure the influence of value preferences through the 'scientization' of policy debate (Nelkin 1975; Habermas 1971; Sarewitz 2004); how expertise is vulnerable to manipulation by policy actors (Nelkin 1975; Collingridge and Reeve 1986; Thorpe 2002; Sarewitz 2004); and how both policy actors and experts attempt to assert policy authority in different issues through either bargaining and negotiation or setting of 'boundaries' between science and politics (Gieryn 1983; Jasanoff 1987, 1990; Bijker *et al.* 2009).

This work thus broadly rejected the traditional assumption that science and politics are wholly autonomous and distinct, and argued they can no longer be thought of as entirely separate inputs to policy. The tendency (if not perhaps the explicit aim) was to dissolve the distinction between political and expert authority by arguing that the latter, rather than deriving from the possession of objective knowledge, is at best politically negotiated and at worst just politics by other means. The worst case is made even more questionable by the accompanying assumption that expert authority represents the values and interests of a minority elite not necessarily concerned with promoting the interests of society as a whole. In its strongest form this argument tended to deny all claims to apolitical knowledge while portraying the production and presentation of science and its evidence as another kind of social activity, or form of theatre, shaped by the politics of time and place and largely focused on the pursuit of power and authority (see, for example, Latour 1987; Hilgartner 2000). The alleged tension between different forms of authority founded on different criteria of legitimacy is thus redescribed as a disguised political contest founded on different sets of interests. Jasanoff (2003: 398) demonstrates this important conceptual shift when she argues that the main issue for STS is 'how particular claims . . . of expertise come into being and are sustained, and what the implications are for truth and justice; the intellectually gripping problem is not how to demarcate expert from lay knowledge or science from politics'. From this perspective, the demarcation between science and politics that occurs in policy making is made, not with reference to the validity of one form of knowledge or legitimacy claim over another, but ultimately through *political* negotiation of the boundary arrangements between actors.

The third wave, according to Collins and Evans (2002), began in tandem with, but also in response to, the constructivist critique of expert authority's role in legitimizing policy. Third-wave debate became largely focused on normative attempts at reconciling the second wave's demolition of Enlightenment-based views of knowledge with the increasing societal demand for more technology and more and better expertise. In effect, this research rediscovered the essential tension between expert authority and democratic authority via its interest in locating expert authority's 'proper place' in contemporary liberal democracy. The basic question of demarcation re-emerged as a problem of finding an appropriate balance between what Kitcher (2011: 20–5) calls 'epistemic equality' and technocratic epistemic elitism. There was a felt need for a means of legitimizing the influence of expert knowledge on policy making in a way that somehow remained consistent with the second wave's trenchant critique of expert legitimacy.

The proposed solutions offered by the so-called 'policy turn' in part reflected an earlier prescription by Collingridge and Reeve (1986) for greater or lesser levels of public participation and oversight depending on a policy issue's level of risk and the potential consequences of experts getting it wrong (see also Funtowicz and Ravetz 1993). Fischer (2009), for example, advocated better democratic oversight of expert influence through greater citizen engagement and deliberative democratic governance. Brown (2009), reflecting the ideas of Latour and Jasanoff, attempted to delineate policy processes that are legitimate, rational and participatory by politicizing expertise in a manner that expanded democratic participation while still reserving an important role in the policy process for scientists and experts. Pielke (2007), on the other hand, argued the need for scientists and experts to act as 'honest brokers' in the policy process, exhorting them to openly declare interests and to avoid acting as policy advocates (see also Bijker *et al.* 2009).

Collins and Evans's third-wave contribution begins by agreeing with wave two that experts do not have privileged access to nature, only a socially mediated path to knowledge. Wave one was thus at fault in conceding excessive authority to 'credentialed' experts while ignoring the policy-relevant tacit skills and experience of non-credentialed players, who ought also be considered experts by virtue of their experience. The 'legitimacy problem' could be solved, they claimed, by 'showing that the basis of technical decision-making can and should be widened beyond the core of certified experts' (Collins and Evans 2002: 237).[5] Nevertheless, Collins and Evans considered that wave two went too far in broadening the net of expertise to include almost everyone, thus making the concept virtually meaningless. Democracy was not best served by attempting to democratize expertise *in toto*. Indeed, if credentialed expertise should know its proper place, then so too should democracy if it wants substantive policy to be informed by specialist knowledge.[6] In short, this middle path, in fact, conceded some foundational, if problematic, validity to a legitimate demarcation between expertise and politics that the second wave had called radically into question.

## The challenge of good public policy

In spite of the more contested status and even suspicion of expert authority among academics and public in the recent times, it is inconceivable in our science-and-technology-dominated world that unelected experts will not continue to play a very large, indeed unavoidable, role in public policy. The ongoing demand for such expertise can be seen in perennial hopes for 'rational' policy and in the still fashionable demand for 'evidence-based' policy (EBP). First championed by the British Labour Party under Tony Blair, EBP is in reality just a reiteration of the Enlightenment-based idea that good public policy must be deeply informed by expert knowledge based on legitimate evidential claims.[7] Its general thrust remains in keeping with first-wave thinking – namely, not to displace political authority with expert knowledge and authority but to resolve or reconcile the two through legitimate ranking: politics will guide and direct while expert knowledge serves and enlightens.

The general state of relations between democratic governments and experts thus remains, despite notorious examples of political friction, one of acute interdependency. Democratic governments (and these days almost *all* governments) are crucially reliant for their continuing legitimacy on providing ever-increasing levels of citizen welfare, which means, in a technological age, fostering and embracing specific forms of knowledge production that might promote welfare through effective policy. And if government is thus dependent on science, science is hardly less dependent on government, which spends enormous amounts of tax income on education, provides tax incentives for research and development in private firms and itself funds direct grants to researchers of every stripe.

Inevitable government dependency on expert knowledge does not imply, needless to say, that evidence and expertise can be simplistically imagined as compelling agreement on policy action. This is partly due to the sheer messiness of the typical policy process. As Kingdon (1995) so aptly notes, in the policy field nothing ever proceeds straightforwardly, and the currents of politics and policy reasoning only seldom flow harmoniously together to provide an agreed solution to a policy problem. But this fluid complexity is compounded by the central irony of the politics-expert relationship – namely, that expert knowledge can maintain its authority and thus its political legitimacy only insofar as it is recognized as the product of genuinely and demonstrably objective processes and analyses, but that in the realm of politics there will always be some with a vested interest in denying that authority by asserting researcher bias, or weakening it on uncertainty grounds. Indeed, the more policy relevant scientists and experts become, the more likely their inputs are to be challenged and discredited in the political arena (Collingridge and Reeve 1986: 4–6).

The ability to challenge the authority of experts is always made easier by the fact that experts, due to the argumentative nature of their disciplines, seldom speak with a unified voice on any issue, particularly on questions of specific detail often asked by policy actors. Variant and even contradictory conclusions are generally the norm in science given the complex nature of reality and the difficulties of decisively penetrating it. This may give ideological opponents the advantage of picking and choosing among expert evidence to suit their predisposed positions on an issue, but there are also undoubtedly cases where genuine differences in beliefs on causation cause good-faith differences on policy prescription resulting in equally vigorous contest.[8]

The interdependency of democratic governments and experts is thus real but often messy, and sometimes conducive to policy stalemates and crises of legitimacy rather than good, broadly supported public policy. Part of the problem is no doubt the growing list of hugely complex, difficult-to-define, 'wicked' policy issues that politics is now expected to manage. Some scholars argue that such issues must inevitably confound traditional Western political institutions, whose effectiveness is highly contingent on the historical and political circumstances in which they were conceived. Hisschemoller (2005), for instance, argues that modern democratic institutions are, by design, good at managing political conflicts over values and interests (which in itself may be questionable) but are ill-suited to dealing with the conflicting knowledge claims that result from attempting to apply rationalist policy approaches to wicked, multi-framed policy problems. Hajer (2003), too, claims that the dynamic nature of modernity and the new kinds of policy problems it inevitably produces (biotechnology, environmental policy, nuclear power, geoengineering) has created 'new political spaces' beyond the scope of classical–modernist political institutions to manage.[9] In such highly complex policy areas, where values are in sharp conflict and uncertainty is pervasive, the border between expert advice and policy advocacy can get very blurred.

The result would seem to be a general 'politicization' of expertise. Sometimes politicization may occur involuntarily through guilt by association – for example, when research is cited to support a political position, or is seen to contradict entrenched values or beliefs (e.g. experts questioning the relative harm of some recreational drugs). At other times experts may become genuine policy entrepreneurs operating as a 'fifth branch' (Jasanoff 1990) or as 'stealth advocates' (Pielke 2007). Brown (2015) argues, however, that politicization is more usefully understood as occurring when expertise or knowledge claims enter a policy environment that has become a site of political contest. Some experts may very well use claims to privileged knowledge to advocate on political questions, while policy makers for their part may try to exclude unwanted advice or set up informal barriers to filter out unpalatable policy assessments. However, these are quite separate issues, according to Brown, from that of how expertise becomes politicized within a contested political process by having its claims willy-nilly aligned with one or another political position. The corollary of this must be that there is much mundane, day-to-day policy making and implementation unaffected by the kind of hot political contests that produce difficult to reconcile tensions between expert and political forms of authority (though how far such tensions are absent even in relatively uncontroversial policy areas would be a matter for close empirical study).

However that may be, we take seriously here the view that the tension between expert and political authority exposed in the literature needs to be further explored and better understood. What emerges from our overview of the three waves is that the identification of this tension is generally accompanied in the scholarship by a powerful impulse to resolve or mitigate it, either by: demarcation and ranking (first wave); redefinition of the two sources as aspects of the 'political' (second wave); or a turn to public supervision of experts (the essence of the 'policy turn') with some also arguing for reconstructed notions of expertise (e.g. Collins and Evans's third-wave contribution).

Missing from these approaches, we argue, is a theoretical willingness to preserve the tension in order to permit close examination of the precise nature of the interaction of these partly competing, partly cooperative, authorities. Our approach seeks to examine the nature of the tensions unencumbered by a concern to rank, dissolve or democratize them. It can perhaps best be described as a meso-level analysis aimed at examining and revealing the nature of the tensions between expert and political forms of authority as they reveal themselves in important public policy cases. We do so to understand how these tensions can be said to contribute to, or undermine, the potential for that good public policy that is a reasonable aspiration of all liberal democracies. Our goal, then, is not to argue for solutions to particular policy problems, or to further elucidate the social underpinnings/constructions of expert authority. Instead, we aim to better understand the dimensions and scope of 'the problem' – that is, how the interaction of expert and political forms of authority complicates the standards required for 'legitimate' policy, or may even alter the very notion of what counts as legitimate policy in liberal democracies.

The illustrative/comparative case studies approach we have adopted will contribute to existing scholarship in two ways: first, by expanding the range of policy areas where tensions have been observably high between expert and political forms of authority; and second by developing a comparative analysis of both the kinds and significance of expert-political tensions that can develop across very different political circumstances and issue areas. Of particular interest is how boundaries between authoritative knowledge claims and values assumed by rationalist policy approaches (e.g. EBP) often become blurred in high stakes policy debates, and also the impact this blurring has on both the nature of the debate (e.g. scientized/politicized) and its prospects for producing 'legitimate' policy decisions.

## The cases

Our approach is comparative because in the very diversity of these significant policy areas we anticipate the potential for gaining general insights regarding political and expert knowledge. It is for this reason that we have selected cases from a broad spectrum of important policy domains where interaction between political and expert authority has produced political conflict and/or ongoing debate. Each chapter explores a series of important questions, including: What are the sources of each form of authority in a case? How does each form of authority manifest itself? What are the relative strengths and weaknesses of both political and expert authority in the case? Does one form of authority dominate, or is there evidence of more balanced interaction? And how is the policy outcome shaped, if at all, by the nature of the interaction between expert and political of authority? Collectively, they reveal the different ways the tension between political and expert authority can become manifest.

The cases are as follows.

Chapter 2, *The Undead Linear Model of Expertise*: Darrin Durant takes up in detail the theoretical issue of the 'three waves' in order to assess the value of preserving the tension (normatively) between expertise and democracy. He argues that resolving the tension between expertise and democracy is flawed either when you defer to experts (wave one) or when you defer to politics (wave two). He uses the case of the collapse of bee colonies to show that wave two can slip back into wave one, but also graduate to wave three. The confusion can be resolved, Durant contends, by not trying to resolve the tension between expertise and democracy.

Chapter 3, *Intelligence and the Use of Armed Force*: Paul Pillar examines the fraught relations between political and expert authorities in security cases where governments resort to armed force, observing the typical strengths and weaknesses of each. He observes that deference to the executive on security issues tends to limit necessary debate on grave matters, while intelligence judgments tend toward an unwarranted narrowing of range upon another nation's weapons programs. Political authority tends toward gut reactions that confuse what is truly political with what is better left to experts, while expert authority is unable to address the important questions about the wisdom of military actions.

Chapter 4, *The Long Goodbye: Science and Policy Making in the International Whaling Commission*: Michael Heazle addresses the International Whaling Commission (IWC), the first international wildlife management regime to explicitly make science the exclusive basis for policy making, to demonstrate how expert authority's influence can wax and wane under changing political priorities and circumstances and even become largely irrelevant under intense and sustained political conflict. He argues that where opposing values are both sharply divided and entrenched, as with whaling, attempts at agreement on the authority of experts alone are likely to stall in the face of uncertainty and conflicting political priorities and interests.

Chapter 5, *On the Interdependency of Political Authority and Economic Expertise*: John Kane observes that modern societies have made themselves utterly dependent on the health and stability of the supposedly independent realm of economics, making the experts in the field—economists—the key authorities on whom governments rely for continued growth. He argues, however, that the claims of the dominant strain of neoclassical economics to scientificity, based on an identification of science and modeling, are bogus, and that the utopian mythology of a naturalistically free market at the heart of such models tends to displace democratic political authority and conduce to crises. He argues nevertheless that there is an acute interdependency between recognized economic expertise and the authority of political leaders charged with 'managing the economy', the former being a function of a political consensus around the ideas and prescriptions of a particular set of economists. The authority of neoliberal economists, assured for decades, was undermined by the Global Financial Crisis, but no alternate consensus had developed around any other group, leaving political leaders at sea about how to find a way back to prosperity.

Chapter 6, *Uneasy Expertise: Geoengineering, Social Science, and Democracy in the Anthropocene*: Clare Heyward and Steve Rayner trace the development of radical geoengineering responses to climate change, showing how the technocratic overtones of previous environmental discourses facilitated the advent of geoengineering research as a serious policy option. They point out the asymmetry in attitudes toward geoengineering and more conventional responses to climate change and suggest that the critical attitudes toward geoengineering are based on particular views of democracy and technological essentialism. Using Cultural Theory, they offer an explanation of why this asymmetry has arisen.

Chapter 7, *Democratic and Expert Authority in Public and Environmental Health Policy*: David Kriebel and Daniel Sarewitz examine how expert and democratic decision-making intersect in surprising and important ways around the problem of regulating toxic chemical use in the United States. Formal governance approaches have produced lengthy regulatory debates but little progress toward reducing production of and exposure to chemicals. Meanwhile, a different model of addressing health risks from synthetic chemicals significantly bypasses expert risk assessment to build alliances between public interest groups and corporations, amalgamating the health and safety community's desire for action with the corporate world's preference for voluntary action. The role of experts remains crucial but is nonetheless secondary to the democratically driven arrangements that civil society and corporations negotiate.

Chapter 8, *In Search of Certainty: How Political Authority and Scientific Authority Interact in Japan's Nuclear Restart Process*: Paul Scalise illustrates how political elites in post-war Japan habitually deferred to scientific authority in regard to Japan's energy challenges and, after the Fukushima Nuclear Power Plant disaster, how they turned again to expert authority for answers regarding nuclear safety, environmental friendliness, and consumer welfare. He shows how science-based, technical judgments of the Nuclear Regulatory Authority (NRA) interacted with the socioeconomic concerns of political authority in the decision-making process. He contrasts the differing responses of political authority depending on whether expert authority was unanimous or divided on existing dangers, or alternatively where it concluded no safety issue existed. His chapter highlights how variation in *perceived* safety and certainty regarding the existence of active fault lines beneath nuclear reactors predetermined the level of political involvement.

Chapter 9, *Drifting to New Worlds: On Politics and Science in Modern Biotechnology*: Haig Patapan explores the contest between political and scientific expert authority by examining developments in assisted reproductive technology, genetically modified foods, stem cell and cloning. He argues that although politics and industry significantly shape and influence biotechnology, scientific innovations are in turn transforming political life—ranging from the meaning of family relations, to the core question of what it means to be 'human'. These incidental, unintended, incremental and far-reaching changes, he claims, radically undermine political authority and are difficult to resist or politically control due to the benefits accruing from the innovations.

Obviously, such a diversity of cases affords abundant material for reflection. In the concluding chapter, *A Democratic Tension?*, Michael Heazle and John Kane attempt to discern individual insights from each study and draw general conclusions for the whole field.

## Notes

1 Witness the Chinese Communist Party's desperate efforts to maintain traditional political authority following Deng Xiaoping's sacrifice of ideology for more pragmatic economic policies and outcomes, and the party's now crucial dependency on expert economic judgments on 'what works' for its ongoing legitimacy. Deng's famous 1962 statement that 'It doesn't matter whether a cat is white or black, as long as it catches mice' serves as an example of his long held preference for 'what works' over ideology. Deng's more pragmatic approach to economic and social policy later led to him being purged from the party twice during the Cultural Revolution (1966–1976; see Vogel 2011).

2 One might argue that the tension between expert and political authority exists 'by design' in liberal democracies, making the exact nature of their joint contribution to legitimate policy always a question of balance. But according to Jasanoff (1990: 8) any such balance, even if achieved, can 'disintegrate any moment as a result of changes in either knowledge or politics'. Whether a satisfactory balance can be consistently maintained across a broad range of complex and contested policy issues in the modern policy environment seems doubtful.

3 Churchill was here quoting from the diaries of his father, Randolph (Churchill 1964: 127).

4 In fact, what scientists and experts have experienced is but part of a diminishment of respect for all forms of alleged authority in modern democracies. The shift is evident in the replacement of the Enlightenment model of an expertly produced *Encyclopedia* by the popularly authored *Wikipedia*. A number of high-profile scientists have not helped the cause by their overreaching. Hughes (2012: 32) describes a willingness among some scientists to accept 'without question the hype that suggests that an advanced degree in some area of natural science confers the ability to pontificate wisely on any and all subjects'. But their interventions are unlikely to have much effect in a society where people in white lab coats, doctors or lawyers are no longer accorded the automatic respect and even reverence they once received.

5 Their attempt to deal with this 'problem of extension', as they call it, relates to what Kitcher (2011: 20–5) has more recently termed the 'epistemic division of labour', which concerns the question of how different sources of knowledge, including those from outside normally recognized sources of scientific and technical expertise, can be recognized and included in the policy process.

6 We are deeply indebted to Darrin Durant for helpful observations on Collins and Evans.

7 EBP seems therefore less a new approach to policy making than a case of pouring old wine into new bottles. Its longevity as a policy model, however, has had the benefit of stoking considerable scholarly concern over the authority, credibility, and proper role of experts in generating and presenting evidence in policy debates. Indeed, the important question for most scholars has not been whether policy should be based on evidence but rather the more difficult questions of *which* evidence should be used, and whose interpretation of it should be invoked to support a particular policy and why.

8 We are indebted to Dan Sarewitz for making this point.

9 The result, in Hajer's view, has been the growth of an 'institutional void' in which the traditional sources of authority and power that drove agenda setting and policy making in the twentieth century have become diffused across a variety of non-traditional policy actors (NGOs, special interest groups, consumer advocates, and so on) and often beyond the borders of the nation state.

## References

Beck, U. 1992. 'From Industrial Society to the Risk Society: Questions of Survival, Social Structure and Ecological Enlightenment'. *Theory, Culture & Society*, 9(1): 97–123.

Bijker, W.E., Bal, R. and Hendriks, R. 2009. *The Paradox of Scientific Authority: The Role of Scientific Advice in Democracies*. Cambridge, MA: MIT Press.

Brown, M.B. 2009. *Science in Democracy: Expertise, Institutions, and Representation*. Cambridge, MA: MIT Press.

Brown, M.B. 2015. 'Politicizing Science: Conceptions of Politics in Science and Technology Studies.' *Social Studies of Science*, 45(1): 3–30.

Churchill, R.S. 1964. *Twenty-One Years*. London: Wiedenfield & Nicolson.

Collingridge, D. and Reeve, C. 1986. *Science Speaks to Power: The Role of Experts in Policy Making*. London: Pinter.

Collins, H.M. and Evans, R. 2002. 'The Third Wave of Science Studies: Studies of Expertise and Experience.' *Social Studies of Science*, 32(2): 235–96.

Collins, H.M. and Evans, R. 2003. 'King Canute Meets the Beach Boys: Responses to the Third Wave.' *Social Studies of Science*, 33(3): 435–52.

Ezrahi, Y. 1990. *The Descent of Icarus*. Cambridge, MA: Harvard University Press.

Feyerabend, P.K. 1976. 'On the Critique of Scientific Reasoning.' *Essays in Memory of Imre Lakatos*. R.S. Cohen, P.K. Feyerabend and M.W. Wartofsky (eds). Dordrecht: D. Reidel Publishing Company, 109–43.

Fischer, F. 1990. *Technocracy and the Politics of Expertise*. Newbury Park: Sage Publications.

Fischer, F. 2009. *Democracy and Expertise: Reorienting Policy Inquiry*. Oxford: Oxford University Press.

Fischer, F. and Forester, J. (eds). 1993. *The Argumentative Turn in Policy Analysis and Planning*. Durham, NC: Duke University Press.

Funtowicz, S. and Ravetz, J.R. 1993. 'Science for the Post-Normal Age.' *Futures*, 25: 739–55.

Gieryn, T.F. 1983. 'Boundary-Work and the Demarcation of Science from Non-Science: Strains and Interests in Professional Ideologies of Scientists.' *American Sociological Review*, 48(6): 781–95.

Habermas, J. 1971. *Toward a Rational Society*. London: Heinemann Educational Books.

Hajer, M. 2003. 'Policy Without Polity? Policy Analysis and the Institutional Void.' *Policy Science*, 36(2): 175–95.

Hilgartner, S. 2000. *Science on Stage: Expert Advice as Public Drama*. Stanford, CA: Stanford University Press.

Hisschemoller, M. 2005. 'Participation as Knowledge Production and the Limits of Democracy.' In *Democratization of Expertise? Exploring Novel Forms of Scientific Advice in Political Decision-Making*. S. Maasen and P. Weingart (eds). Dordrecht: Springer.
Hoppe, R. 2005. 'Rethinking the Science-Policy Nexus: From Knowledge Utilization and Science Technology Studies to Types of Boundary Arrangements.' *Poiesis Prax*, 3(3): 199–215.
House of Lords. 2000. 'Science and Technology—Third Report.' Retrieved at www.publications.parliament.uk/pa/ld199900/ldselect/ldsctech/38/3801.htm
Hughes, A.L. 2012. 'The Folly of Scientism.' *The New Atlantis*, fall, 37: 32–50.
Jasanoff, S. 1987. 'Contested Boundaries in Policy-Relevant Science.' *Social Studies of Science*, 17(2): 195–230.
Jasanoff, S. 1990. *The Fifth Branch: Science Advisers as Policymakers*. Cambridge, MA: Harvard University Press.
Jasanoff, S. 2003. 'Breaking the Waves in Science Studies: Comment on H.M. Collins and Robert Evans, "The Third Wave of Science Studies".' *Social Studies of Science*, 33(3): 389–400.
Jasanoff, S. (ed.) 2004. *States of Knowledge: The Co-Production of Science and Social Order*. London: Routledge.
Kane, J.W. and Patapan, H. 2012. *The Democratic Leader*. Oxford: Oxford University Press.
Kingdon, J. 1995. *Agendas, Alternatives and Public Policies*. New York: HarperCollins.
Kitcher, P. 2011. *Science in a Democratic Society*. New York: Prometheus Books.
Kuhn, T. 1996. *The Structure of Scientific Revolutions*, 3rd ed. Chicago: University of Chicago Press.
Lane, R.E. 1966. 'The Decline of Politics and Ideology in a Knowledgeable Society.' *American Sociological Review*, 31(5): 649–62.
Larvor, B. 1998. *Lakatos: An Introduction*. New York: Routledge.
Latour, B. 1987. *Science in Action*. Cambridge, MA: Harvard University Press.
Lindblom, C.E. 1959. 'The Science of "Muddling Through".' *Public Administration Review*, 19(2): 79–88.
Lindblom, C.E. 1979. 'Still Muddling, Not Yet Through.' *Public Administration Review*, 39(6): 517–26.
McClay, W.M. 2009. 'What Do Experts Know?' *National Affairs*, 1: 145–59.
March, J. and Olsen, J. 1989. *Rediscovering Institutions: The Organizational Basis of Politics*. New York: Free Press.
Nelkin, D. 1975. 'The Political Impact of Technical Expertise.' *Social Studies of Science*, 5(1): 35–54.
Pielke Jr, R., 2007. *The Honest Broker: Making Sense of Science in Policy and Politics*. New York: Cambridge University Press.
Popper, K.R.1972. *The Logic of Scientific Discovery*. London: Hutchinson.
Ravetz, J.R. 1971. *Scientific Knowledge and Its Social Problems*. Oxford: Clarendon Press.
Sarewitz, D. 2004. 'How Science Makes Environmental Controversies Worse.' *Environmental Science & Policy*, 7(5): 385–403.
Thorpe, C. 2002. 'Disciplining Experts: Scientific Authority and Liberal Democracy in the Oppenheimer Case.' *Social Studies of Science*, 32(4): 525–62.
Uphoff, N. 1989. 'Distinguishing Power, Authority and Legitimacy: Taking Max Weber at His Word by Using Resources-Exchange Analysis.' *Polity*, 22(2): 295–322.
Vogel, E.F. 2011. *Deng Xiaoping and the Transformation of China*. Cambridge, MA: Belknap Press.

Wynne, B. 1996. 'May the Sheep Safely Graze? A Reflexive View of the Expert-Lay Knowledge Divide.' In *Risk, Environment and Modernity*. S. Lash, B. Szerszynski and B. Wynne (eds). London: Sage, 44–83.

Wynne, B. [2002] 2003. 'Seasick on the Third Wave? Subverting the Hegemony of Propositionalism: Response to Collins & Evans.' *Social Studies of Science*, 33(3): 401–17.

# 2 The undead linear model of expertise

*Darrin Durant*

Sometimes the claims of natural science about reality are both dead and alive; they are 'undead', as Bart Simon (1999) showed in documenting the continued research on the phenomena of cold fusion long after the controversy was supposedly closed in 1990. Sometimes social science claims about the world are also undead – for instance, technological determinism as a model of social change is *passé* but enduring (Wyatt 2008), and the deficit model of public engagement with science is both categorically repudiated and continually reinvented (Wynne 2006). Joining this list of walkers among the undead is the linear model of expertise, which holds that the science-policy relationship ought to be one where we get the facts right and then act. The model suggests a decision procedure where truth speaks to power, with science assuming the role of arbiter of preferences and meanings and prescriber of policy, on the assumption that expertise reduces uncertainties and more expertise equates to better policy. Despite Science and Technology Studies (STS) and other cognate fields dealing the linear model of expertise innumerable mortal blows (cf. Jasanoff and Wynne 1998; Beck 2011), the model groans on. The purpose of this chapter is to suggest this might not be such a bad thing. Preserving aspects of the linear model of expertise implies a preservation of the tension between expertise and democracy, but I suggest that tension is crucial for any democracy aiming to retain the possibility that specialist knowledge can inform deliberation.

The problem the linear model of expertise is aimed at solving is the tension between expertise and democracy. Guston (1993: 18) has referred to the tension as an essential one based in 'the exclusionary and authoritarian nature of science and the egalitarian requirements of democracy'. Turner (2001, 2003) notes a similar tension but suggests that it can be handled by normal democratic checks and balances if conceived of as asymmetries in access to knowledge or violations of conditions of State equality, but is a serious issue when manifested as discretionary power wielded inside bureaucracies with weak external accountability structures. Collins and Evans (2002, 2003, 2007) generated controversy when they suggested that we could break into three waves the responses to the tensions between expertise and democracy. In wave one, expert authority is thought to resolve public policy disputes. The path to resolving tensions between expertise and democracy, by treating expertise as potentially apolitical and value free, turns on letting experts decide or letting experts frame what counts as the common good. In wave two, the special authority of expertise is deconstructed. The path to resolving tensions between expertise and democracy, by treating expertise as politics by other means, turns on extending the public franchise and choosing your politics. Collins and Evans proposed wave three, which borrows wave two's social account of expertise but is not allergic to demarcations and boundaries like wave two. Wave three branches out from wave two in three ways. First, wave three restricts ordinary publics to the political phases of disputes, in part by invoking a distinction between the roles of intrinsic and extrinsic politics in technical assessment. Second, wave three reconstructs expertise along realist lines, treating expertise not as just an attributed property resulting from social conflict but as an acquired possession resulting from social immersion and experience in a domain of practice. Third, where wave two seems to allow that politics should completely rule expertise, wave three specifies a two-way relation. The stronger the degree of technical consensus the more should political actors feel a constraining force, even though politics retains both priority and the right to overrule expertise.

My point in this chapter is a relatively simple one: the balancing motif of wave three is more preferable to the leveling motif of wave two. To make this case the chapter unfolds in three more sections. In section two I unpack Collins and Evans's three waves, showing their point to be that complete deference to experts (wave one) is just as normatively inadequate as complete deference to publics or politics (wave two). In section three I thus take a closer look at calls to democratize expertise. Drawing on recent analysis of the debate over colony collapse disorder among bees, I show that a wave two approach can devolve to a wave one approach. Things begin well with an impulse to level the epistemological playing field, but begin to go awry when a normative rebalancing of political influence is suggested. The political values recruited as guides to suitable change can sit uneasily next to straight implications that one side has the science right and policy would be wiser if it followed that side's preferences. When that happens, the linear model of expertise is reinvented in the implicit devolution to a 'speaking truth to power' model. In section four I conclude by advancing a speculative position. Without what Collins *et al.* (2010: 197) call a 're-constructed concept of expertise' we might lose touch with the ability of expertise to act as a check on raw political and economic power. If we want a balance of expert and political input, we should resist dissolving expertise into politics. Preserving a tension between expertise and democracy—which owes more to balancing than leveling—might result in productivity for democracy rather than its undermining.

## Three waves

Sensible people agree that we cannot resolve tensions between expertise and democracy by leaving matters of concern to experts. Expert results and expert framings where they feed into politics fail most tests of democratic legitimacy, and often of epistemic diversity too. However, nor can we resolve the tension between expertise and democracy by relying on infinitely extending the public franchise and simply choosing our politics. Collins and Evans (2002: 237) referred to the issue as one of the 'problem of extension': how far does public participation go before we lose a sense of the value and proper role of expertise altogether?

A reader who is familiar with wave two deconstructions of expertise might object here that few have explicitly called for a complete ignoring of expertise. But that is my point. We have a gap between rhetoric and reality. When not criticizing expertise as a scientistic incursion into democracy, wave two often leaves vague the role expertise might actually play in public policy, or only parenthetically and grudgingly reintroduces a role for expertise if challenged about the apparent omission. Wave two deconstructions of science, which have informed some of my own work critical of nuclear industry claims of technical virtuosity (cf. Durant 2009a, 2009b), are correct but incomplete. The analyst's rush to establish cultural credibility as one of the cool kids endorsing possibly unconstrained public participation leaves expertise unreconstructed after its deconstruction. Collins and Evans address this gap, seeking to reconstruct expertise, thereby making a contribution to preserving a useful tension in our democracies where knowledge can inform power.

## *Oh, that linear model of expertise*

The need for Collins and Evans to reconstruct expertise arises in part from the politically unpalatable way expertise is couched in the linear model of expertise. That model itself grows out of the broader linear model of innovation: autonomous scientific communities generate basic research, which leads by a sequential process through applied research and development to production and diffusion, the output ultimately taking the form of economic growth and social benefits. Not surprisingly, the linear model of innovation is said to have either never existed outside straw man functions (Edgerton 2005), or to be a rhetorical entity sustained by successive communities of natural scientists, business schools and economists that 'is still alive despite. . . a proclaimed death' because of the seductive power of statistics (Godin 2006: 641). The linear model of expertise has a family resemblance relationship with the general idea embedded in the linear model of innovation, of the necessity of getting the facts right, with the idea that better policy eventuates being a subcategory of broad social benefits.

Echoing the challenge Edgerton (2005) and Godin (2006) issued to those discussing the linear model of innovation, we could at this point go on the academic equivalent of a bug hunt and seek out exemplary instances of authors articulating a linear model of expertise. However, Hounshell's (2005) response to Edgerton is quicker and possibly more informative. Hounshell argued, based on an examination of the practices of the chemical giant DuPont, that without explicitly articulating or even advocating a linear model of expertise, the thinking and practices of decision-makers at the company were completely consistent with the model. In their discussion of the linear model Jasanoff and Wynne (1998) proceed in a similar fashion, finding the approach of the Intergovernmental Panel on Climate Change (IPCC) undermined by its adherence to a linear 'speaking truth to power' model. Science apparently closes gaps in knowledge and eliminates uncertainties, allowing policy making to then take over. Such linear model assumptions are not exhaustively documented, like a jury model of textual prosecution. Instead, the argumentative style is that the IPCC's approach is consistent with the linear model of expertise. Jasanoff and Wynne's (1998: 16) reflexive point is that 'social and cultural commitments are built into every phase of knowledge production and consequent social action, even though enormously effective steps are often taken to eliminate the traces of the social from the scientific world'.

Criticisms of a posited or assumed linear model of expertise thus typically proceed by targeting particular assumptions about the foundations of expert knowledge and the relationship between experts and policy fields. As Lovbrand and Oberg (2005: 195) note:

> the idea of a linear relationship between science and policy where scientific knowledge functions as the rational basis for decision-making builds upon realist epistemology in which science, either through unbiased observation or universal reason, gains access to true representations of reality.

As anyone familiar with a few decades' worth of constructivist analysis of science would know, such realist assumptions appear weak when confronted by detailed empirical study of science in practice. Facts and values are revealed as intertwined, politics and expertise as rolled up in each other, and uncertainty and contingency appear constitutive of science. As Jasanoff (1990: 250) once said of the hope that expertise could be some kind of neutral bastion of truth arbitrating disputes, 'agencies and experts alike should renounce the naïve vision of neutral advisory bodies "speaking truth to power," for in regulatory science, more even than in research science, there can be no perfect, objectively verifiable truth'.

Conceptions of how facts are to be put to use have also been revised, for the idea that facts compel actions exaggerates how policy making works. The idea that facts compel actions also 'derives from the linear model of expertise, in which the interaction between science and politics is conceived of as unidimensional, linear, and one-way: from science to policy ("truth speaks to power")' (Beck 2011: 298). Beck further notes that the assumption of linearity is supported by three propositions: more and better scientific research creates certainty, resolves political disagreements, and results in an evidence-based policy process replacing raw political contestation. None of these propositions stand up to detailed scrutiny. Sarewitz (2004) has shown that more and better research can often make controversies worse, because science supplies contesting parties with competing facts, and discipline-based understandings of issues can be causally tied to competing value-political positions. Similarly, Stehr and Grundmann (2011) note that the application of science via expertise is necessarily open to disagreement due to expertise being demand driven. Most concur that science does not produce absolute certainty, only situated judgments of reliability and plausibility (Collins and Evans 2003: 435–6; Evans and Plows 2007: 833–4; Jasanoff and Wynne 1998: 76; Collins 2009: 39; Stehr and Grundman 2011: 105–6). Science can thus become the battleground for political dispute rather the grounds to resolve political dispute (Pielke 2005; Stehr and Grundman 2012: 147). As Sarewitz (2000) argues, science becomes the battleground for politics because of a strength not a weakness: the ability to saturate politics with facts and thus provide resources for partisan battles. As science feeds into the political process it can thus result in a scientized politics. What ought to be open debate about values and politics is either subsumed under a debate about the quality of science or displaced to the realm of science with partisan scientific claims standing as proxy for value-based positions (Pielke 2005). Ultimately, the pursuit of political ends occurs while combatants simultaneously proclaim to be disengaged from politics (Jasanoff 2008).

## A third wave as the next step?

Collins and Evans (2002) joined this discussion by periodizing into waves the efforts to resolve the tensions between expertise and democracy. Wave one presented the policy process as capable of smooth operations if it deferred to apolitical and certain science, thereby displaying little awareness of the importance of democratic legitimacy in decision-making. Wave two deconstructed science, showing it to be imbued with politics. The route to democratic legitimacy is presented as one going through public participation. Wave three codifies a worry that wave two illuminated the problem of legitimacy but created the problem of extension: how far should public participation go?

Collins and Evans had found themselves agreeing both with the claim that science reflects social interests, assumptions and frameworks, and that there must be some kind of gap between experts and lay publics if the idea of respecting the skills, training and experience of experts had any meaning (Collins, Weinel and Evans 2010: 185–6). The relationship between the waves was specifically articulated (Collins and Evans 2002: 240; 2003: 437–40; Collins *et al.*, 2010: 185–7; Collins 2013: 159, 170–1). While wave two showed wave one to be bankrupt, wave three should run parallel to wave two, sharing its constructivist outlook but adding a new (realist) rationale for why expertise should be valued. The aim is to avoid collapsing expertise into politics. Moreover, wave three shares with wave two a critique of scientistic ideas – for instance, the idea that the only sound answers are scientific answers, or that only the propositional questions of experts are the legitimate way to frame a debate.

Lampooning wave one is not new, but Collins and Evans courted controversy by suggesting that wave two also opted for an unsatisfactory resolution to the tensions between expertise and democracy. Echoing John Dewey's quip from 1927 that 'the cure for the ailments of democracy is more democracy' (1981: 327), wave two therapy involves empowering citizens to provide input on technical matters, shape the meanings and issue definitions at stake in public policy, and frame decision-making and deliberation in a broadly civic, not narrowly technical fashion. Collins and Evans worried that a justifiable critique of scientism can morph into an anti-science position in which propositional questions have no place in debate (2003: 438). They similarly worried that public participation—premised on a leveling of the epistemological field—can devolve into a bland populism insensitive to trying to theorize domains in which the judgments of those who know what they are talking about can carry more weight than non-experts, even if not apolitically certain along (mythological) wave one lines (Collins *et al.* 2010: 187).

Wave three thus denies that 'ordinary members of the public should be full parties to technical debates' (Collins 2013: 174). But Collins and Evans also have a reconstructed conception of expertise, simultaneously expanding it and contracting it, so the restrictions on publics does not have the same force as it would coming from a wave one perspective. Expertise is not defined by the possession of formal qualifications, but by the possession of experience and tacit knowledge. Narrow bands of publics can thus acquire expertise via their situated experience, but otherwise the folk wisdom view that ordinary publics can make the same kind of sound judgments about technical matters as specialist experts is rejected. Wave three, in effect, reconstructs expertise along realist not attributional lines. Expertise is not just whatever happens to emerge from passages of socio-scientific conflict. Expertise is a possession of specialists, acquired via social immersion in a community and experience in a problem domain, embodied or embrained as skills and judgmental abilities, and guided by formative intentions somewhat captured by a set of neo-Mertonian values. Rather than reinvent the fact-value distinction, the aim is to provide a value-value distinction (Evans and Plows 2007: 833; Collins *et al.* 2010: 191–2; Collins 2013: 172). Expertise operates according to a set of values or formative intentions, broadly Mertonian in their emphasis on universalism, disinterestedness, and openness to criticism, but also supplemented by specifying why the locus of legitimate interpretation is probably best to be closer to the producers of knowledge. Lay public participation is thus to be limited to 'domains where no specialist expertise is required' (Collins *et al.* 2010: 187), specifically discussed as the political phases of disputes.

Collins and Evans's reconstruction of expertise carves out a domain for experts via a series of demarcations: between technical and political phases of debates (Collins and Evans 2002: 261–8, 282–3, 286; Evans and Plows 2007: 833–6; Collins *et al.* 2010: 192–4), and between intrinsic and extrinsic politics (Collins and Evans 2002: 245, 2003: 441; Collins *et al.* 2011: 342–3; Collins 2013: 165). The technical phase hinges on resolving propositional questions and matters of fact, presenting a picture of the state of play with regard to research in a specialist problem domain, and drawing on those who know what they are talking about to arrive at the best decision even if in the long term it turns out to not be the correct decision. Here the compatibility of wave three with the political philosophy of John Rawls is apparent (cf. Durant 2011). As Collins (2013: 165) noted, citing Rawls, 'the method of agreement' is the standard 'rather than the agreement itself'. The political phase is concerned with the preferences and interests and bargaining that inform public deliberation and policy making. Moreover, while politics is intrinsic to science, meaning it cannot be eliminated only managed, this should be distinguished from extrinsic politics. In the political phase extrinsic or overt politics is legitimate, but in the technical phase, technical matters decided by explicit, overt extrinsic politics are considered illegitimate. Put simply, that politics is embedded in science is not the same as willingly endorsing the reaching of politically predetermined conclusions.

Collins and Evans stress that the political phase ultimately has priority over the technical phase of disputes (Collins *et al.* 2010: 188), and can indeed overrule it provided the political grounds are made clear and explicit (Collins 2013: 172). A reconstructed (socialized) linear model of expertise has a home in wave three, but it is carefully circumscribed. Experts are the best source of advice in the technical phases of disputes, and expertise functions as one source of constraint on political processes dominated by either populism or controversies manufactured to serve dominant political–economic interests. The path to resolving tensions between expertise and democracy, by treating our situation as framed by the need to make the best decision despite the flawed character of expertise, turns on choosing the values we wish to inform expertise: an outlook Collins and Evans label 'elective modernism' (Collins 2009, 2013; Collins *et al.* 2010).

### *The empire strikes back*

Many critics of Collins and Evans virtually ignored the reconstructed concept of expertise populating the newly created technical phase. Instead, the critics focused on the perceived slight to publics, arguing that the global move to embrace public participation meant that Collins and Evans's restrictions on publics were akin to trying to 'lock the barn door after the horse has already bolted' (Jasanoff 2003: 397).

Yet Jasanoff's quip merely demonstrated her own blindness, as if asking where the horse is headed is a silly question. Jasanoff overlooked the way in which, amid our global rush to involve publics in deliberation and decision-making, there is a simultaneous discussion of appropriate limits. It is certainly true that public participation is advocated as a key ingredient to resolving policy disputes involving expertise, a commitment STS shares with many other disciplines, including political philosophy (cf. Gutmann and Thompson 2004), risk assessment (cf. Otway 1987), and environmental sociology (Bulkeley and Mol 2003). There is a broad acknowledgement that a (pure) linear model of expertise both flaunts norms of democratic legitimacy and fails at solving the problems it might be aimed at solving (cf. Braun and Kropp 2010). But the sentiment Collins and Evans were worried about was nicely captured by Seyla Benhabib (1996: 67) who quite reasonably argued that democratic legitimacy flows from 'free and unconstrained deliberation of all about matters of common concern'. Unconstrained? As is well known in political theory, various ideas captured by the generic label 'deliberative democracy' actually admit of constraints on deliberation. Benhabib (1999: 412) herself admitted the need for constraints, opposing the project of securing group rights to the extent that doing so might encourage 'the tyranny of intolerant minorities and narcissistic collectivities'. Fabre (2003: 107) thus notes that deliberative democrats 'disagree on the kind of reasons citizens can advance, on what the common good is, and on which political procedures best capture the deliberative ideal'.

Jasanoff and other critics of Collins and Evans often write as if any restriction on publics is undemocratic, suggesting an uninformed or least narrowly parochial grasp of democratic theory (cf. Durant 2010). As I have argued elsewhere (Durant 2011), different models of democracy exist, and Collins and Evans's position has much in common with the position of John Rawls, while Jasanoff and many others in STS deploy positions much closer to that of Jurgen Habermas. Neither of those giants of political philosophy endorses *unconstrained* public participation. Each thinks it important to theorize limits on publics and limits to the scope of politics as a contribution to discussing where civil society is heading. In the next section, I analyze some recent (wave two) work purporting to show the weaknesses involved in Collins and Evans's restrictions on publics and the virtues of democratizing expertise. Unfortunately, the case actually illustrates a weakness in wave two, because if interpreted strictly, it devolves to wave one, although if interpreted loosely it is actually an instance of wave three.

## Dead bees

Sainath Suryanarayanan and Daniel Lee Kleinman (2011, 2012, 2013a, 2013b, 2014) have recently published a series of outstanding articles on the case of colony collapse disorder among bees. Colloquially referred to as 'mad bee disease', bee colony die-offs have been observed since the mid-1990s. As a crucial insect pollinator in industrial agriculture, bee die-offs impact the agricultural, economic and environmental health of a host of nation-states. Pathogens, parasites, pesticides and malnutrition have all been blamed, along with the possibility that a complex cocktail of factors is responsible rather than a single factor. The controversy also involves debates about the role of regulatory regimes in setting appropriate standards for assessing laboratory and field-based evidence. Indeed, the proper epistemic base for deciding the issue is in dispute, with academic and industrial toxicologists typically facing off against commercial beekeepers over competing conceptualizations of causes. Ultimately, this case is about the democratization of expertise, with calls to pay more attention to the claims of commercial beekeepers and to have a full democratic debate about the values that ought to inform regulatory standards. Thus, Suryanarayanan and Kleinman recommend that regulators 'need to seriously value a much broader array of knowledge forms, practices, and actors' (2011: 35), that we ought to pay attention to 'alternative knowledge claims promoted by less powerful stakeholders' (2012: 494), 'explore the place of "non-scientist" stakeholders' (2013a: 216) and aim for a 'more genuinely participatory research' (2013b).

***Setting up some sides***

Suryanarayanan and Kleinman set up their story in classic fashion, pitting two sides against each other: toxicologists versus commercial beekeepers (cf. 2012, 2013a). Or that battle transposed to a battle between non-precautionary and precautionary regulatory approaches (cf. 2011, 2013b). Or the whole issue amplified into a battle wherein toxicologists in the US defeat commercial bee-keepers aided by the US non-precautionary regulatory regime, and commercial beekeepers in France defeat toxicologists aided by the French precautionary regulatory regime (cf. 2014). Suryanarayanan and Kleinman also create academic sides. The editors of the volume in which the Suryanarayanan and Kleinman (2014) chapter appears note that Suryanarayanan and Kleinman's case study 'is an example of delivery on the deferred promise of Collins' third phase' (Hess and Frickel 2014: 17). This is in reference to the 'third phase' of Collins's 1980s empirical program of relativism, which called for an examination of the way in which wider social relations structure scientific assessment (and note Collins specifies he has done much work in wave two mode of showing how expertise is socially situated). But Suryanarayanan and Kleinman echo Jasanoff's (2003: 390) critique of Collins and Evans, that compared to Collins's third phase, the third wave of Collins and Evans seems a 'drawing in of the horns'.

Collins and Evans's wave three approach of categorizing experts, phases of disputes and types of politics is labeled by Suryanarayanan and Kleinman a 'definitional or typological exercise' that 'fails to address an important set of issues' (2013a: 218). Such typological categorizations are criticized as too context independent to address historicized processes of knowledge organization and dominance creation. While Collins and Evans specify conditions under which non-certified experts have legitimate claims to be heard, Suryanarayanan and Kleinman (2013a) think acts of 'calling for' inclusion fail to 'consider the factors that legitimize certain claims' (2013a: 219), or explain how 'expertise becomes delegitimized' (2013a: 232). Indeed, Suryanarayanan and Kleinman indict Collins and Evans for failing to understand that 'possession of "contributory expertise" does not guarantee recognition' (2013a: 218) and that 'entitlement does not guarantee influence' (2013a: 233). It takes very little political theory to realize that something is awry with Suryanarayanan and Kleinman's critique, once we ask whether recognition and influence ought to be guaranteed. This strong demand appears innocent of the classic distinction between two forms of equality: *ex ante* (equality in people's prospects) and *ex post* (equality in people's outcomes). Without parcelling out what they actually expect reforms of expert-public relations to accomplish, such as equalizing opportunities or going much further to legislate equal outcomes, Suryanarayanan and Kleinman appear to fall into the trap Ronald Dworkin (1987) exposed – that is, equality of opportunity is a fine ideal, but equality of outcome might demand not just too much practically, but too much ideally. Implicit ideals of equality of influence potentially conflict with other egalitarian ideals like autonomy, the tailoring of one's resources, and the benefits of persuasion via mutual reflection.

The problem with Suryanarayanan and Kleinman's position is that its animating normative principle owes more to the metaphor of leveling than balancing. Thus the nub of their dispute with Collins and Evans, say Suryanarayanan and Kleinman, is that the politics of expertise is not played out on a 'symmetrical field, but rather in a complex structure of relations of domination and subordination' (2013a: 219). Rather than ride Collins and Evans's third wave, Suryanarayanan and Kleinman advertise their approach as applying the work of Jasanoff and Wynne, specifically the focus on how 'knowledge constitution . . . does not occur on a symmetric field' and how processes of 'imposition' and 'marginalization' occur (2013a: 219). Obviously, facts get shaped in asymmetric fields of influence. The question is: should this fact be addressed by leveling or balancing? Leveling is a rhetorically appealing ideal, drawing on our sympathies with ideals of equality, but in practice often amounts to an implied rebalancing of a field of influence toward some preferred set of social actors. In the next two subsections I will first show how Suryanarayanan and Kleinman level the epistemological playing field, but then show how they actually seek a rebalancing toward the claims of commercial beekeepers. I suggest they end up sliding from their avowed critique of scientism into a scientistic speaking-truth-to-power approach, via an uneasy mix of denigrating one set of technically disfavored experts while championing the policy implications that flow from a politically favored set of experts.

### The case

Using Suryanarayanan and Kleinman (2013a) as the primary articulation of their case, their aim is to explain the 'asymmetrically large influence of toxicologists' (2013a: 218) and show how 'credibility' and 'influence' in disputes over knowledge 'is shaped by the historically established social organization of knowledge production. Certain forms of knowledge production—*epistemic forms*—gain credibility over time and become institutionalized' (2013a: 219; emphasis in original). According to Suryanarayanan and Kleinman, toxicologists 'conceptualize' Colony Collapse Disorder (CCD) as a 'complex, multifactorial phenomenon' but their 'experimental research is animated by' efforts to 'isolate individual factors and their direct, causal roles' (2013a: 229). The epistemic form of the toxicologists is thus described as favoring formal procedures and measures, and a causally driven experimental approach, tending toward a focus on single factorial, rapid, and lethal effects (2013a: 226–7). By contrast, while the 'rhetorical form' (2013a: 223) of commercial beekeepers sometimes 'stresses the elimination and isolation of individual causal factors' (2013a: 222) and the community of commercial beekeepers is admitted to be 'highly heterogeneous', Suryanarayanan and Kleinman focus on commercial beekeepers who attribute the cause of CCD to the use of 'neonicotinoids', a systemic pesticide (2013a: 221). The epistemic form of the commercial beekeepers is described as favoring informal procedures and measures and a field-based approach, tending toward a focus on synergistic, sublethal and cumulative effects (2013a: 222–3).

Furthermore, there is a large difference in the respective communities over which type of error warrants more efforts to avoid. A false-positive or type I error means that you establish an effect where in fact there is none, and a false-negative or type II error is where one does not establish an effect while there is actually an effect. Suryanarayanan and Kleinman (2013a) describe toxicologists as having a 'preference for erring towards false-negative conclusions [type II errors], where a substance that is harmful is incorrectly concluded to be safe' (2013a: 230). Basically, toxicologists want to avoid making type I errors (claiming an effect that does not exist) and so are more tolerant of type II errors (missing an effect that does exist). That false-negative orientation reflects a 'nonprecautionary "sound science" approach toward pesticide regulation' (p. 231). By contrast, commercial beekeepers are described as having a preference for erring toward 'arriving at false-positive conclusions (type I errors)' (2013a: 223), where a substance that is harmless is incorrectly concluded to be unsafe. Basically, commercial beekeepers want to avoid making type II errors (missing an effect that does exist) and so are more tolerant of type I errors (claiming an effect that does not exist). That false-positive orientation reflects the 'precautionary principle' of prohibiting use of a substance in the 'absence of evidence that such substances are safe' (2013a: 223). Put colloquially, toxicologists are avoiding crying wolf whereas commercial beekeepers prefer false alarms.

Suryanarayanan and Kleinman (2013a) note that neither form of expertise is apolitical. The epistemic form of the toxicologists is explained as related to the primacy and growth of agricultural and entomological lab research organizations and the agro-economic contexts within which early entomologists worked. Academic toxicologists are said to prefer the agro-entomological approach due to their career stakes, goals of enhancing their cultural capital and desire to achieve intellectual distinction. In short, Suryanarayanan and Kleinman trace a 'convergence' (2013a: 232). The formal and quantitative methods favored by academic honey bee scientists were mirrored by the US regulator—the Environmental Protection Agency (EPA)—shifting from a precautionary to non-precautionary approach. This shift was in part a response to agrochemical manufacturers pushing for a definitive evidence evidentiary standard. The scientist/regulator convergence means that those agrochemical companies willingly support such regulatory standards, as demonstrating adverse effects from their products becomes very challenging.

Suryanarayanan and Kleinman (2013a) explain the epistemic form of the commercial beekeepers as an instance of ordinary folk struggling with competing drivers that shape their activities, such as commercial beekeepers' practical experiences, commercial interests, livelihoods at stake, and historical tensions with farmers that were themselves shaped by the political economy of the US pollination market (2013a: 223–4, 232). Commercial beekeeping moved from predominantly honey production *pre*-1940s to increasingly serve the task of managed pollination of crops in industrial agriculture settings *post*-1940s (2013a: 220–1), setting up ongoing farmer/beekeeper tensions. While farmers could increase productivity and beekeepers could increase revenue, farmer short-term productivity goals are not always in step with beekeeper long-term concerns to sustain bee colony health. For instance, commercial beekeeping has incorporated the use of antibiotics and pesticides to maximize pollination potential and control disease, but these practices have led to concerns about bee health (especially with regard to reproduction, malnutrition and pests). A recent shift in industrial agriculture (2013a: 224–5) has seen older insecticides (their mode of action being direct contact and short lived) replaced by newer systemic insecticides (toxins accumulate and remain active and persistent in treated plants for long periods of time). The EPA is said to understand these systemic insecticides—such as neonicotinoids—as possessing more insect toxicity but less risk to humans than traditional insecticides.

### *From leveling to rebalancing*

Suryanarayanan and Kleinman's (2013a) story has leveled the discursive field: we have two different kinds of science backed by two different regulatory principles, leaving us suspicious about toxicologist's 'dominance' and commercial beekeeper's lack of 'legitimacy and influence' (2013a: 233). The false-negative orientation of toxicologists and their allies in the EPA and industry is labeled 'not a transparently appropriate means for measuring the effects of honey bee exposure to pesticides' (2013a: 230). Commercial beekeepers are depicted as getting an unfair hearing, with their expertise dismissed as anecdotal, just not data, trial and error or ad hoc (2013a: 225). The credibility of toxicologists is dragged down and the credibility of commercial beekeepers is pushed upwards, leveling off the competitors and leaving the reader wondering what accounts for such dominance and dismissal.

Trouble begins, though, when the rebalancing starts. Tackling the question of why the beekeepers were ignored starts off reasonably well. Suryanarayanan and Kleinman (2013a) dismiss the potential explanation that 'agrochemical industry actors have direct economic power' (2013a: 225) and opt instead for an explanation stressing 'indirect dimensions of epistemic dominance' (2013a: 226). It is 'a matter of social history' (2013a: 226), reflecting a 'convergence in the epistemic orientations of academic toxicologists and the EPA' (2013a: 232) that resulted in 'less direct, less explicit, and less strategic actions across a substantial historical expanse' (2013a: 230). Suryanarayanan and Kleinman claim the primacy of toxicologists' epistemic form is not due to the 'inherent superiority' of their evidence (2013a: 233), but to structural factors shaping the social organization of knowledge. Elsewhere Suryanarayanan and Kleinman (2012) use the same case to show how the epistemic form of the toxicologists institutionalizes particular kinds of ignorance about colony collapse disorder (2012: 498, 508) and justifies regulatory inaction (2012: 494). Again, the epistemic form of the toxicologists is said to reflect a particular amalgam of history, culture, norms and interests rather than 'inherent superiority' (2012: 509, 497).

Is that same constructivist—shall we say wave two—approach to factual knowledge extended to commercial beekeepers? No, because Suryanarayanan and

Kleinman imply that some superior knowledge is being overlooked in overlooking the commercial beekeeper knowledge. The dominance of the epistemic form of the toxicologists is said to be liable to produce 'false knowledge' or 'factual ignorance', with toxicologist's knowledge labeled 'poorly representative' and liable to 'distort subsequent knowledge' (2012: 508–9). Field experiments of toxicologists are based on an 'implausible assumption' (2013a: 229). Toxicologists' expertise is persistently criticized for being little more than 'artificially reductive experimental designs' (2011: 36), 'misleading' and prone to 'overlook' effects (2012: 508–9). Indeed, the epistemic form of toxicologists is ultimately dismissed because it 'does not embody the settings in which CCD actually arose' (2012: 510). Here the reader's presumed lack of sympathy for the hegemonic actor wielding too much influence is drawn on to mask a straight claim about who is getting it right. Is it not inconsistent to deny inherent superiority to one test but imply some other test can take us all the way to the real world? Suryanarayanan and Kleinman attempt to avoid the inconsistency by specifying 'our point is not to say that commercial beekeepers always know best' (2013b) and by acknowledging that 'all research requires choices and thus has biases' (2011: 36). However, the disclaimers ring hollow once we directly compare the presentation of the toxicologist's technical claims with that of the commercial beekeepers.

The knowledge of commercial beekeepers is said to derive from an 'on-the-ground epistemic form' reflecting bee hives 'that exist in real world contexts' (Suryanarayanan and Kleinman 2013a: 223). The authors also state that commercial beekeepers deal in 'real-world environmental complexity' (2011: 36) and 'on-the-ground evidence' (2014: 104). Because commercial beekeepers have 'methodically studied' (2014: 103) colony collapse disorder and carried out 'careful and systematic observation . . . we should take their research seriously' (2013b). Maybe, but note the revealing contrast: toxicologist's knowledge needs to be seen as 'matters of social history, not nature' (Suryanarayanan and Kleinman 2011: 34). Notice the zero-sum form? This 'social, not natural' claim is either an unfortunate rhetorical excess or suggestive of deep asymmetry. Recall David Bloor's Strong Programme specified that symmetry implies both society and nature will be implicated, in a non-zero-sum fashion, in all institutionalized forms of knowledge (1999: 87–91). The kind of asymmetric demarcations criticized by Bloor, often symbolized by the scare quotes used to label the (not really) 'scientific' views of dissenters, opponents or minority views, actually return in Suryanarayanan and Kleinman's zeal to identify those on the wrong side of the truth. For instance, when some commercial beekeepers are found who do not agree with their colleagues that neonicotinoids are responsible for honey bee deaths, they are dismissed in classic demarcation fashion as 'dissenting "scientific" beekeepers' (2014: 104). Suryanarayanan and Kleinman appear to believe themselves in possession of some metric for how to measure closeness to nature, given the distance they open up between toxicologists and nature, a distance closed down when talk turns to commercial beekeepers.

Suryanarayanan and Kleinman write of commercial beekeepers' regulatory approach in similar 'they know best' terms. Toxicologists' preference for erring toward false-negative conclusions is said to be 'misguided' (2011: 33) and not 'transparently appropriate' (2013a: 230). By contrast, the example of the commercial beekeepers is said to point to the features of what would be a good regulatory regime, one favoring a preference for false-positive over false-negative conclusions (2011: 35; 2013a: 230), one accepting of 'suggestive data' not just 'unambiguous proof' (2011: 35), and one having a precautionary orientation (2011; 2013b). Indeed, the contextual and multifactorial correlations approach favored by the commercial beekeepers points to a research regime that would be interdisciplinary (2012: 511) and even transdisciplinary (2011: 36). The implied contrast is to some stodgy old disciplinary approach consigning bees to early death. Ultimately, for Suryanarayanan and Kleinman, the precautionary policy approach *should* be our 'default position'; certainly, they are 'on the side of those who prefer to err on the side of caution' (2013b). Note here that Collins and Evans (2002: 239) claimed that 'scientism' in wave one often involved social analysts seeking to reinforce the success of the sciences, as well as social analysts regarding good science as putting the analyst in a position to speak with authority and decisiveness. Suryanarayanan and Kleinman's advocacy for a marginalized perspective wins them good guy's status, but they are nevertheless treating the knowledge of commercial beekeepers as good science that authorizes policy inferences.

The key here is to avoid the temptation to give Suryanarayanan and Kleinman what could be called 'the constructivist's free pass', overlooking their denigration of toxicologist's knowledge on the grounds that the constructivist is, of course, impartial about truth. That may or may not be so, on a case-by-case basis, but the strong contrast Suryanarayanan and Kleinman draw between toxicologist and commercial beekeeper knowledge sets up the science practiced by commercial beekeepers as the better science. The classic linear model of expertise is not reproduced because neither side is presented as possessing apolitical knowledge. Instead, the linear model of expertise is reinvented for a constructivist age, the focus now on finding kinds of science that match pre-existing political sensibilities. In this case, it is precautionary regulatory ideals, but note the immediate difficulty that unless you are an ideologue the precautionary principle alone cannot select out for you all of the expertise(s) you will favor.

For instance—and although it is an argument for another day—pushing the precautionary principle to its limit might mean the suggestive evidence marshalled by anti-vaccinationists, opponents of anthropogenic global warming, and proponents of the hormesis hypothesis that low doses of radiation are good for you, would all be instances of neglected expertise that our policy worlds need to act on more than they currently do. Of course, in some nations this is already true. The principle of charity would suggest that Suryanarayanan and Kleinman are aware of such possible counter-arguments: you save some bees, but you might endanger the kids, the climate, and anyone trying to protect themselves from radiation hazards. The inherent merit of the precautionary principle cannot carry the full weight of Suryanarayanan and Kleinman's policy recommendations, implying that their policy recommendations are ultimately not separable from an endorsement of particular knowledge claims. Once we realize that the merit of some scientific research is thus closely linked to a policy recommendation, we can see how the linear model of expertise is being reinvented even if not reproduced.

Aside from the assumptions of apolitical science and automatic resolution of political disagreements, the linear model of expertise is back, reinvented as some marginal knowledge claims' ability to alert us to facts (pesticide hazards) demanding our policy response. Using the language of science policy interfaces discussed by van der Sluijs (2012), Suryanarayanan and Kleinman appear to be endorsing 'working deliberatively within imperfections' but come much closer to 'speaking truth to power'. A key difference between the two interfaces, suggests van der Sluijs, is that while the former embraces uncertainties as a fact of life, the latter tends to over-emphasize the uncertainties of positions being criticized. Suryanarayanan and Kleinman end up looking like the conservative critics of climate change analyzed by van der Sluijs, basing (proposed) policy on marginal knowledge claims while ridiculing consensual science.

## Concluding speculations

The linear model of expertise was recently described as 'the most widespread the world over' (Polk and Kain 2015: 8), presumably because it is 'deeply embedded in both the science and the policy communities' (Sarewitz 2000: 81) and reflects dominant cultural assumptions and implicit framings (Jasanoff and Wynne 1998: 3). What should we do about this undead visitor, refusing to die despite so many mortal blows? Put differently, what should we do about experts continuing to exert power? The 'living dangerously' reply is to live with it. Though I expect that reply to be taken out of context by those with participatory axes to grind, there is an analogy here with the welfare debate, where it is said that the more you tighten entitlement rules to restrict welfare fraud, the more you risk cutting off citizens who truly need assistance. Similarly, the more we subject expertise to the conceptual death of completely blurring it into politics, the more we risk strangling the ability of experts to tell us when uninformed populists and/or powerful political and economic elites might have little we want to call scientific adorning their claims.

So is there a way to love the old linear model of expertise without getting bitten? Think of the linear model of expertise as having a core assumption, that facts compel policy (strong version) or that expertise can/should speak 'truth to power' (weak version). That core assumption is flanked on one side by epistemological assumptions about apolitical facts and the possibility of the truest science, and on the other side by political assumptions about facts reducing disagreements and wise policy capable of being directly inferred from facts. Suryanarayanan and Kleinman's mad bee disease case is an excellent wave two analysis that illustrates what is wrong with each of the flanking assumptions, but it unfortunately rests on an uneasy mix of arguments underlying the shift from leveling the epistemological field to rebalancing the political field. However, that does not mean that Suryanarayanan and Kleinman bring into disrepute the core assumption, certainly not in the weak version, and possibly not even in the strong version. Let me explain.

In their wave two mode, have Suryanarayanan and Kleinman done much more than what Sarewitz (2000, 2004) criticized as the recruiting of particular expert claims as partisans to a political outlook? As Pielke (2005: 47) termed it, treating 'science as a proxy for political battle'. Recall that commercial beekeeper knowledge was presented as a truer knowledge than that of toxicologists, and thus worthy of informing and in some sense limiting policy. This seems a slide back into the wave one scientism Suryanarayanan and Kleinman critique, as we have a version of the epistemological side of the flanking assumption noted above. But reflecting the uneasy mix of arguments bedeviling their normativity, the precautionary principle acted as an ethical-political guiding light pointing to which truth should be speaking to power. Suryanarayanan and Kleinman (2011) tie the appropriateness of that political choice to ideals of inclusiveness, epistemic diversity, and health and environmental hazard reduction. Yet it is not transparently obvious that operating with suggestiveness and tolerance for type I errors as an evidentiary standard is always the most democratic or technically satisfactory way of proceeding.' For starters, it might not be as balanced as it sounds. Freudenburg *et al.* (2008: 29–31) suggest that while focusing on type I errors while ignoring type II errors is potentially insensitive to public and environmental health, a single-minded obsession with either type I or type II errors reflects a lack of balance because of the way type I and type II errors trade off against each other. Put more speculatively, if avoiding type I errors means that you avoid too easily assuming a proposition is true, it takes little thought to see how that kind of standard might impact on nuclear industry claims that low doses of radiation are good for you, or anti-vaccinationists' claims that autism results from the measles, mumps and rubella shot, or claims by opponents of anthropogenic global warming that natural cycles suffice as an explanation of warming trends.

Here an ironic twist emerges. Is there anything inherently wrong with experts speaking their truth to power? Despite the uneasy mix involved in getting there, Suryanarayanan and Kleinman have identified an important political phenomenon, of a group of experts (who happen to be marginalized) suggesting that one policy direction poses potential hazards to public and environmental health and that another policy direction might better guard against such risks. Expertise is entering the political process as not just another voice, but as a potential source for setting the grounds of political debate: what Hess (2014) calls 'doxological power'. Hess notes that traditional science-politics models granted to expertise such doxological power: the capacity to establish 'the underlying area of the uncontested that is created through the engagement of orthodoxies and heterodoxies', which 'limits but does not determine the myriad battles over what *should be* the case' (2014: 128). Although Suryanarayanan and Kleinman's ostensibly wave two analysis deconstructs toxicological expertise, an image nevertheless emerges of commercial beekeeper expertise as reliable enough to guide policy. Indeed an image emerges of expertise as flawed but enviable, based in skills and judgments related to experience and participation in a community of practice, political but not thereby necessarily bought and sold, uncertain but not thereby politically otiose. Suryanarayanan and Kleinman also intimately associate the knowledge of the toxicologists with the agrochemical industry: such knowledge 'served the companies that produce the newer systemic insecticides' (2013a: 232) and 'benefits the short term interests of agrochemical producers' (2011: 36). Ironically, we are not too far from Collins and Evans's call for a reconstructed conceptualization of expertise. Moreover, are we seriously to believe that lurking in the background of the overall negative appraisal of toxicologist's knowledge is not something like Collins and Evans's warning against extrinsic interests guiding science? Suryanarayanan and Kleinman either doth protest too much or simply misunderstand what they are criticizing.

The ultimate source of Suryanarayanan and Kleinman's troubles is that, having deconstructed expertise but refused to reconstruct expertise, they are forced to slip a preferred set of expert claims in through the political back door as a preference for a particular regulatory ideal. Suryanarayanan and Kleinman thus grant policy power to their preferred expertise in a clumsy fashion, via persistently subtle pernicious contrasts aimed at suggesting commercial beekeepers have got it right. Collins and Evans took the honest path of theorizing how to value expertise in the political process even when you know expertise is not 'god-like'. In response they have to rebut accusations of technocracy (Collins and Evans 2003: 439–40; Collins *et al.* 2011: 344), even though they explicitly deny that the role of experts has to hinge on settling who is right (Collins 2013: 171). Clumsiness aside, Suryanarayanan and Kleinman thus share with Collins and Evans a commitment to the ideal that expertise can exercise doxological power. A caveat would be that Collins and Evans would apply conditions to the 'limit', such as degrees of consensus and any prohibitions falling mostly on political actors' freedom to distort technical consensus, but I suspect most would agree such caveats to be wise. Nevertheless, for all the fancy arguments about expertise blurring into politics, sometimes we have to look for what is *lamented*. Hess (2014) thus lamented the epistemic rift that has allowed policy about climate change to be detached from consensual science. Suryanarayanan and Kleinman lament the fact that commercial beekeepers in the US lack doxological power (their French cousins having achieved it).

Collins and Evans have voiced similar laments about epistemic rifts between policy fields and consensual science (in cases involving anti-vaccinationists, AIDS denialists, and opponents of anthropogenic warming, cf. Collins *et al.* 2010, 2011; Collins 2013). The difference is that wave three gets to the laments via the demarcations that wave two is allergic to but often implicitly presupposes: explicit value judgments should not determine results in the technical phase, and that where they do not, the technical phase can play a role informing politics and limiting though not determining policy choices. Doxological power is catered for and, while politics can ignore expertise, the third wave shares with the work of Sarewitz and Pielke an emphasis on explicitness being a virtue when political and value judgments are made to ignore or overrule or apply judgments arrived at by experts. Expertise may not provide unambiguous answers to policy dilemmas nor compel action, but what happens if we give up on the doxological power of expertise and completely decouple or insulate politics from scientific advice?

Freudenburg *et al.* (2008) hint at the result. They discuss the history of what they call SCAMs (Scientific Certainty Argumentation Methods): the tactic of demanding absolute scientific certainty before regulatory action can proceed, which, if successful, can mean the safety of a practice or technology might never even arise (2008: 27). Although SCAMs appear effective even against the best science (2008: 17), as seen in cases including pesticide regulation, global warming, tobacco, and lead in gasoline, an underlying message of their article is the dangers that occur when the political process can be captured by powerful vested interests intent on rejecting any limit to their actions. Science can obviously be ignored, but that can be to the detriment of public and environmental health, just as abuses of doxological power can be detrimental. Experts wielding doxological power in an accountable fashion might just be to contribute to an essential tension in democracy, where the economically, politically and culturally powerful can have their discretionary authority limited. Alvin Gouldner (1973: 80–1) once said that to reconstruct society means to reconstruct theory. Our modern task might be to reconstruct our notions of expertise, and how it feeds into policy, with a view to not losing touch with the doxological power of expertise despite its human frailties. Collins and Evans's wave three seems a good start to that project.

And in what way is that project a liberal democratic project? Wave three rejects the idea that being subject to power is necessarily demeaning, a revision to the liberal canon by questioning whether liberalism takes for granted a concept of adulthood in which dependency is inherently shameful, and whether that is a sustainable model of respect in a world of divisions of labor and inequalities (cf. Sennett 2003: 103). Or consider that Lukes's (2005: Chapters 2 and 3) revision of his 1974 conception of the third dimension of power noted that power as domination, or 'power over', neglects the ways power over others can be productive, transformative and compatible with dignity (2005: 109). The doxological power of experts can obviously be power as domination, but to limit doxological power to power as domination might be to ignore the ways that expert claims can be productive, transformative and dignity-enhancing. Or just a check on unbridled populism and/or political and economic jerks. As Collins and Evans (2007: 8) put it, 'democracy cannot dominate every domain—that would destroy expertise—and expertise cannot dominate every domain—that would destroy democracy'.

## References

Beck, S. 2011. 'Moving Beyond The Linear Model of Expertise? IPCC and the Test of Adaptation.' *Regional Environmental Change*, 11(2): 297–306.

Benhabib, S. 1996. 'Toward a Deliberative Model of Democratic Legitimacy.' In *Democracy and Difference: Contesting the Boundaries of the Political*. Seyla Benhabib (ed.). Princeton, NJ: Princeton University Press, 67–94.

Benhabib, S. 1999. 'The Liberal Imagination and the Four Dogmas of Multiculturalism.' *The Yale Journal of Criticism*, 12(2): 401–13.

Bloor, D. 1999. 'Anti-Latour.' *Studies in the History and Philosophy of Science*, 30(1): 81–112.

Braun, K. and Kropp, C. 2010. 'Beyond Speaking Truth? Institutional Responses to Uncertainty in Scientific Governance.' *Science, Technology, & Human Values*, 35(6): 771–82.

Bulkeley, H. and Mol, A.P.J. 2003. 'Participation and Environmental Governance: Consensus, Ambivalence and Debate.' *Environmental Values*, 12(2): 143–54.

Collins, H. 2009. 'We Cannot Live by Scepticism Alone.' *Nature*, 458(March 5): 30–31.

Collins, H. 2013. 'The Role of the Sociologist After Half-a-Century Studying Science.' *South African Review of Sociology*, 44(1): 158–79.

Collins, H. and Evans, R. 2002. 'The Third Wave of Science Studies: Studies of Expertise and Experience.' *Social Studies of Science*, 32(2): 235–96.

Collins, H. and Evans, R. 2003. 'King Canute Meets the Beach Boys.' *Social Studies of Science*, 33(3): 435–52.

Collins, H. and Evans, R. 2007. *Rethinking Expertise*. Chicago: University of Chicago Press.

Collins, H., Weinel, M. and Evans, R. 2010. 'The Politics and Policy of the Third Wave: New Technologies and Society.' *Critical Policy Studies*, 4(2): 185–201.

Collins, H., Weinel, M. and Evans, R. 2011. 'Object and Shadow: Responses to the CPS Critiques of Collins, Weinel and Evans' "Politics and Policy of the Third Wave".' *Critical Policy Studies*, 5(3): 340–248.

Dewey, J. 1981. 'The Public and its Problems.' In *The Later Works of John Dewey, 1925–1953*, vol. 2. J.A. Boydston (ed.). Carbondale, IL: Southern Illinois University Press, 235–372.

Durant, D. 2009a. 'Responsible Action and Nuclear Waste Disposal.' *Technology in Society*, 31(2): 150–7.

Durant, D. 2009b. 'Radwaste in Canada: A Political-Economy of Uncertainty.' *Journal of Risk Research*, 12(7/8): 897–919.

Durant, D. 2010. 'Public Participation in the Making of Science Policy.' *Perspectives on Science*, 18(2): 189–225.

Durant, D. 2011. 'Models of Democracy in Social Studies of Science.' *Social Studies of Science*, 41(5): 691–714.

Dworkin, R. 1987. 'What is Equality? Part 4: Political Equality.' *University of San Francisco Law Review*, 22: 1–30.

Edgerton, D. 2005. '"The Linear Model" Did not Exist: Reflections on the History and Historiography of Science and Research in Industry in the Twentieth Century.' In *The Science-Industry Nexus: History, Policy, Implications*. K. Grandin, N. Wormbs, and S. Widmalm (eds). New York: Watson, 31–57.

Evans, R. and Plows, A. 2007 'Listening Without Prejudice? Re-discovering the Value of the Disinterested Citizen.' *Social Studies of Science*, 37(6): 827–53.

Fabre, C. 2003. 'To Deliberate or Discourse. Is *That* the Question?' *European Journal of Political Theory*, 2(1): 107–15.

Freudenburg, W.R., Gramling, R. and Davidson, D.J. 2008. 'Scientific Certainty Argumentation Methods (SCAMs).' *Sociological Inquiry*, 78(1): 2–38.

Godin, B. 2006. 'The Linear Model of Innovation: The Historical Construction of an Analytical Framework.' *Science, Technology & Human Values*, 31(6): 639–67.

Gouldner, A. 1973. 'Remembrance and Renewal in Sociology.' In *For Sociology*. New York: Basic Books, 68–81.

Guston, D. 1993. 'The Essential Tension in Science and Democracy.' *Social Epistemology*, 7(1): 3–23.

Gutmann, A. and Thompson, D. 2004. *Why Deliberative Democracy?* Princeton, NJ: Princeton University Press.

Hess, D.J. 2014. 'When Green Became Blue: Epistemic Rift and the Corralling of Climate Science.' In *Fields of Knowledge: Science, Politics and Publics in the Neoliberal Age. Political Power and Social Theory*, vol. 27. S. Frickel and D.J. Hess (eds). Bingley: Emerald, 123–53.

Hess, D.J. and Frickel, S. 2014. 'Introduction: Fields of Knowledge and Theory Traditions in the Sociology of Knowledge,' in *Fields of Knowledge: Science, Politics and Publics in the Neoliberal Age*, vol. 27. S. Frickel and D.J. Hess (eds). Bingley: Emerald Books, 1–30.

Hounshell, D. 2005. 'Industrial Research: Commentary.' In *The Science-Industry Nexus: History, Policy, Implications*. K. Grandin, N. Wormbs and S. Widmalm (eds). New York: Watson, 59–65.

Jasanoff, S. 1990. *The Fifth Branch: Science Advisers as Policymakers*. Cambridge, MA: Harvard University Press.

Jasanoff, S. 2003. 'Breaking the Waves in Science Studies.' *Social Studies of Science*, 33(3): 389–400.

Jasanoff, S. 2008. 'Speaking Honestly to Power.' *Scientific American*, 6(3): 240.

Jasanoff, S. and Wynne, B. 1998. 'Science and Decisionmaking.' In *Human Choice and Climate Change, Volume 1: The Societal Framework*. S. Raynor and E.L. Malone (eds). Columbus, OH: Battelle Press, 1–86.

Lovbrand, E. and Oberg, G. 2005. 'Comment on "How Science Makes Environmental Controversies Worse" by Daniel Sarewitz. *Environmental Science and Policy*, 7: 385–403 and "When Scientists Politicise Science: Making Sense of the Controversy over the Skeptical Environmentalist" by Roger A. Pielke Jr' *Environmental Science and Policy*, 7: 405–17.' *Environmental Science and Policy*, 8(2): 195–7.

Lukes, S. 2005 [1974]. *Power: A Radical View*, 2nd edn. London: Macmillan.

Otway, H. 1987. 'Experts, Risk Communication, and Democracy.' *Risk Analysis*, 7(2): 125–9.

Pielke Jr, R. 2005. *The Honest Broker: Making Sense of Science in Policy and Politics*. Cambridge: Cambridge University Press.

Polk, M. and Kain, J-H. 2015. 'Co-Producing Knowledge For Sustainable Urban Futures.' In *Co-producing Knowledge for Sustainable Cities: Joining Forces for Change*. M. Polk (ed.). New York: Routledge, 1–22.

Sarewitz, D. 2000. 'Science and Environmental Policy: An Excess of Objectivity.' In *Earth Matters: The Earth Sciences, Philosophy, and the Claims of Community*. R. Frodeman (ed.). Upper Saddle River, NJ: Prentice Hall, 79–98.

Sarewitz, D. 2004. 'How Science Makes Environmental Controversies Worse.' *Environmental Science & Policy*, 7(5): 385–403.

Sennett, R. 2003. *Respect in a World of Inequality*. New York: W.W. Norton.

Simon, B. 1999. 'Undead Science: Making Sense of Cold Fusion after the (Arti)Fact.' *Social Studies of Science*, 29(1): 61–85.

Stehr, N. and Grundmann, R. 2011. *Experts: The Knowledge and Power of Expertise*. New York: Routledge.

Stehr, N. and Grundmann, R. 2012. *The Power of Scientific Knowledge: From Research to Public Policy*. Cambridge: University of Cambridge Press.

Suryanarayanan, S. and Kleinman, D.L. 2011. 'Disappearing Bees and Reluctant Regulators.' *Issues in Science and Technology*, 27(4): 31–6.

Suryanarayanan, S. and Kleinman, D.L. 2012. 'Dying Bees and the Social Production of Ignorance.' *Science, Technology & Human Values*, 38(4): 492–517.
Suryanarayanan, S. and Kleinman, D.L. 2013a. 'Be(e)coming Experts: The Controversy Over Insecticides in the Honey Bee Colony Collapse Disorder.' *Social Studies of Science* 43(2): 215–40.
Suryanarayanan, S. and Kleinman, D.L. 2013b. 'Honey Bees Under Threat: A Political Pollinator Crisis.' *The Guardian*, May 8. Retrieved at: www.theguardian.com/science/political-science/2013/may/08/honey-bees-threat-political-pollinator-crisis
Suryanarayanan, S. and Kleinman, D.L. 2014. 'Beekeepers' Collective Resistance and the Politics of Pesticide Regulation in France and the United States.' In *Fields of Knowledge: Science, Politics and Publics in the Neoliberal Age*, vol. 27. S. Frickel and D.J. Hess (eds). Bingley: Emerald Books, 89–122.
Turner, S. 2001. 'What Is the Problem with Experts?' *Social Studies of Science*, 31(1): 123–49.
Turner, S. 2003. *Liberal Democracy 3.0*. London: SAGE.
Van der Sluijs, J. P. 2012 'Uncertainty and Dissent in Climate Risk Assessment: A Post-Normal Perspective.' *Nature and Culture*, 7(2): 174–95.
Wyatt, S. 2008. 'Technological Determinism Is Dead; Long Live Technological Determinism.' In *Handbook of Science and Technology Studies*, 3rd edn. E.J. Hackett, O. Amsterdamska, M. Lynch and J. Wajcman (eds). Cambridge, MA: MIT Press, 165–80.
Wynne, B. 2006. 'Public Engagement as Means of Restoring Trust in Science? Hitting the Notes, but Missing the Music.' *Community Genetics*, 9(3): 211–20.

# 3 Intelligence and the use of armed force

*Paul R. Pillar*

The application of intelligence to decisions about the use of armed force resembles in some respects the interaction of political and expert authority on other issues of public policy but also exhibits some important differences. One of those differences is that the expert authority in question operates largely in a secret world. The output of that authority, in the form of information collected from intelligence sources and analysis contained in classified reports, is only partially revealed to the outside world. Only over time is the output apt to become known slowly and incompletely, sometimes after mistaken impressions and speculation about its role already have been formed. Decision-making about issues of war and peace, not only the intelligence input to decisions, may be clothed in secrecy maintained in the name of national security; this constitutes a further impediment to understanding the respective roles played by different forms of authority.

Whatever influence secret intelligence has on public policy, decisions may be exerted through either of two channels. One is the direct shaping of the perceptions and thinking of the decision-makers. This channel is contained largely within the secret world. Its opacity is a function of the closed doors behind which some of the most important deliberations about the use of armed force are made, and of the difficulty of reconstructing objectively the thought processes of the decision-makers to determine which influences really shaped their perceptions of the foreign situations with which they were dealing.

The other channel is through public opinion, insofar as it influences decisions of war and peace. Public opinion may set informal but generally acknowledged limits on what a government can do, or alternatively push a government into doing what it may otherwise be disinclined to do. The influence of public opinion might also be felt through popularly elected legislatures that authorize, or refuse authorization for, the use of armed force. Either way, the moving of intelligence from its customary secret world into public discourse becomes an important variable. Exactly which intelligence moves across that line, how it is moved, and how the publicly revealed intelligence is interpreted or spun can become major issues or points of controversy, in addition to however this movement affects the ultimate decision about the use of armed force.

Another distinctive characteristic of decisions involving the use of armed force is that they are more political than many other matters of public policy to which expert opinion is relevant. A choice between war and peace is one of the highest profile policy decisions a government can make, and thus it is more likely than other topics to engage entire political systems. The general public is more aware of the issue, and politicians are more expected to develop firm and well-founded views on it. The issue is likely to be politicized in multiple senses of that word.

Such an issue is also highly political in that it hinges on some questions on which political leaders are apt to consider themselves at least as expert as anyone else. Although sometimes seemingly more technical questions are also involved, such as the progress a foreign country is making in developing certain types of weapons, usually the policies, intentions, and motivations of foreign political leaders are the key questions. And who, one might ask, is better qualified to assess those than another political leader? This dimension is in addition to the inherently political, non-technical questions of how a prospective use of armed force fits into a nation's larger objectives and grand strategy and whether the particular objective to be pursued through armed force is worth the human and material costs involved. In any competition between political and expert authorities, the politicians start on this topic with major advantages.

## Nature of political authority

National security is at the core of the most important functions and responsibilities of any political authority. Physical protection of a nation from foreign threats is usually given precedence over almost every other national priority or function of government. Issues of war and peace tend to be viewed in this way even when they do not entail defense against a direct threat to the national homeland. Once armed force is involved or potentially involved, the label 'national security' tends to get slapped on the matter, and national security tends to get equated with being of highest national importance. It is as the custodians of the nation's security that political leaders derive much of their authority.

The power to use military force is usually concentrated in the executive branch of government. Typically, the chain of command for a nation's armed forces runs up to the most senior political executive, although in some systems with a separate head of state and head of government, the top of that chain may be a president while a prime minister is the foremost policy maker. Even in countries such as the United States, with a separation of powers between executive and legislative branches of government, there is much deference to the executive on matters involving the use of military force. The constitutional power of the US Congress to declare war has fallen into disuse, with Congressional resolutions authorizing the use of military force being only a partial substitute.

The political executive's authority regarding the use of armed force manifests itself in two ways. One is the direct command of military resources. This is the aspect of authority that matters most with relatively small military endeavors that do not become a subject of broad national debate. The political leadership can simply issue an order for the armed forces to do something, and they do it.

On the larger uses of military force that do engage the entire political system and become matters of national debate and even controversy, another aspect of political authority is more important than formal command of the armed forces. This aspect is the role of national political leaders as persuaders, appealing to the public and to members of the legislature to support whatever armed endeavor is at issue. It is the role that US president Theodore Roosevelt once referred to as the bully pulpit. The function is one of salesman rather than commander. Success in closing a sale with the public rests on perception of the top political leader as foremost protector of the nation—as the person whose responsibilities and access to information make him best qualified to decide what is best for the country.

## Nature of expert authority

The claim of intelligence to be an expert authority has two bases, which correspond to the two fundamental parts of intelligence work: collection and analysis.

Intelligence is first of all a collector of information to which no one else in a national debate has first-hand access, or at least that no one else can collect without significant difficulty. Issues do not become matters for intelligence in the first place if the relevant information were readily available through other channels. An intelligence agency can be an authority, in other words, because it knows certain things that no one else does, or at least knows them first before disseminating the information to others. This claim to authority is strongest when the information in question concerns closed societies where the regime most assiduously hides its activities from outsiders. If the issue involves possible use of armed force by liberal democracies, most often the target country is such a closed society. War against another liberal democracy does not tend to be an issue in liberal democracies; that is what the much studied democratic peace theory is about (Russett 1994).

Even when the issue involves a closed police state, the claim of exclusive access to information is never absolute. Outside skeptics always have something to go on, in questioning what intelligence says or judgments supposedly based on intelligence. There are, at a minimum, the public statements and publicly observable actions of the regime in question. Over time the exclusivity of information collected through intelligence methods has eroded further, getting even to more technical matters such as unconventional weapons programs. New technologies have put increased information even about closed societies in the hands of the public. Commercial satellite imagery has improved to the point that any educated outsider who wants to pronounce on what is going on at a military base or nuclear plant, even in a notoriously closed society such as North Korea, at least has a fairly clear image on which to base such pronouncements.

The other ground on which to consider intelligence an expert authority involves the assessment and interpretation of information. Although analysis is just as much a core function of intelligence as collection is, it is inherently more vulnerable to challenge—by politicians and policy makers, or by anyone else. However experienced and expert may be the analysts that an intelligence service has working on a problem, there always will be other experts—in academia, think tanks, or elsewhere—willing and able to offer their assessments of the same problem. Even if the expert credentials of an outsider may be subject to question, opinions are cheap.

Outside analysts will, for the reasons just noted in connection with collection, always have unclassified material to work with. Sometimes they will get access to classified material as well. A prominent instance of this was the 'Team B' exercise in the United States in the 1970s, which involved the assessment of strategic nuclear forces of the Soviet Union. A group of hardline defense intellectuals who believed the US intelligence community was understating the Soviet threat was given access to the same classified reporting as the intelligence analysts and prepared their own, more ominous, assessment (Hessing Cahn 1993: 24–5). Later developments in the Soviet force showed the Team B report to have been more mistaken than the official assessment, but in the meantime it had a discernible effect on public discourse in the United States about the US–Soviet arms race.

The principal manifestations of the expert authority embodied in intelligence are classified documents disseminated within the government. These typically include both 'raw' reports, which are lightly processed versions of the information gathered from human or technical intelligence sources, and 'finished' intelligence, which consists of analysis that incorporates information from selected raw reporting into coherent assessments.

Several possible mechanisms can move intelligence, either raw or finished, into public view. Intelligence services themselves release some unclassified products, most often because they are obligated to do so. The US intelligence community is required by law to submit annually to Congress both an unclassified and a classified version of a comprehensive statement of worldwide threats. Congress from time to time calls on intelligence agencies to provide other unclassified statements and testimony. The intelligence agencies occasionally release unclassified products on their own initiative. On any issue that is at all a matter of public controversy, however, most intelligence officers would prefer to keep themselves and their products out of public view. The type of situation they hope to avoid is exemplified by the release in 2007 of an unclassified version of intelligence judgments about Iran's nuclear program—a release that elicited charges that intelligence officers were attempting to limit policy options for dealing with Iran (Office of the Director of National Intelligence 2007).

Other public releases of intelligence output are the result of ad hoc initiatives by policy makers in an effort to support or sell their policies. Usually this consists of incorporating intelligence derived information in statements or speeches by political leaders. The intelligence agencies that originated the information generally have a right of review to protect their sources and methods. Such review may blur or eliminate some details, but it rarely means a veto over release of the information.

A final avenue for release of classified intelligence is leaking. With broad dissemination of intelligence products throughout the parts of government concerned with national security, leaks can originate in many possible places, and the actual point of origin is seldom found. Any leak is inherently a selective extraction from the corpus of available intelligence, with the selection based on whatever happens to be the leaker's agenda.

## The mystique of intelligence

Notwithstanding the disadvantages for intelligence in any competition for influence with political authorities, the public invocation of intelligence by political leaders themselves in mustering support for their policies demonstrates there must be some offsetting advantages. Political leaders are inconsistent in public references to intelligence, invoking it when they believe it will help their case and leaving it unmentioned when they do not believe it will. But at times the impression is conveyed of basing a case primarily on intelligence findings. A leading example is the public campaign by the governments of the United States and United Kingdom to develop support for the invasion of Iraq in 2003. The campaign was designed to make it appear that a major foundation of the decision to launch the war was intelligence about Iraqi programs to develop unconventional weapons.

This type of campaign has the attraction for policy makers of providing a ready-made scapegoat if the use of armed force in question goes awry and becomes unpopular, as the war in Iraq did. The policy makers in the US administration of George W. Bush subsequently took full advantage of this aspect to divert to the US intelligence community as much blame for the war as possible. The generally low popularity of intelligence agencies in the United States facilitated this blame shifting. Intelligence is publicly invoked as an authority, however, mainly for two more positive aspects of the place that intelligence services commonly occupy in public perceptions.

One concerns the unique access that collectors of secret intelligence have to information that is not generally available to others. Intelligence gets invoked as an authority because intelligence services are presumed to have more information about a topic than anyone else has. The intelligence agencies have access to the same openly available material that everyone else does, in addition to what they alone are able to collect.

The other attraction of invoking intelligence as an authority is that a professional intelligence service has a presumed impartiality and objectivity that political leaders, with their policy agendas and partisan identifications, do not have. This does not mean that intelligence services are not accused of having their own cognitive biases; they often are, especially in the wake of prominent intelligence failures. But insofar as a service is seen to have some standing independent of that of the policy maker, what it says is taken as a separate view—a confirming view, if it agrees with the policy makers' assertions.

Both these attributes of intelligence are pertinent only to selling a policy to the public (or to a popularly elected legislature). They are not reasons to expect intelligence to have any particular influence on the policy makers' own decision to employ armed force, or to seek public or legislative approval to employ it.

The exclusive access to secretly collected information is an advantage that intelligence services have over the public but not over policy makers, who are the services' principal customers and have access to most of the same classified material that intelligence officers do. As for the presumption of impartiality, although this accords with the prescribed role of an intelligence service, even policy makers who see the advantage of incorporating intelligence into a pitch to the public do not necessarily see intelligence officers as unbiased. The architects of the Iraq War in the US administration of George W. Bush instead were highly suspicious of intelligence officers as having policy biases opposed to their own (Pillar 2011: 171).

## Political authority in control

Despite whatever positive mystique intelligence may have in the public mind, political authority has overwhelming advantages that cause it to predominate in any issue involving the use of armed force. The advantages are related to the characteristics, mentioned at the beginning of this chapter, that issues involving the application of armed force usually display.

In the partly secret, partly public manner in which such issues play out, the political authority has the ultimate power to determine what becomes public. Intelligence services routinely have a large measure of control over their own material. When the outcome of a public debate about policy is not at stake, the services determine the limits of dissemination of that material even within the government, sometimes restricting access to especially sensitive items to circles much smaller than all those with high level security clearances. The criteria according to which an intelligence service exercises such control, however, are generally limited to the protection of intelligence sources and methods. The services do not have an independent mandate to release or withhold their material selectively on other grounds, and especially not to favor one policy option over another.

Almost any intelligence material can be modified or redacted to make at least some portion of it safely releasable to the public. The substance or conclusion of an intelligence report, for example, can be provided without mentioning the ultimate source of the information, thus not jeopardizing the source. This means that the political authority, if it sees release of some such information as advantageous in building public support for its favored policy, can direct the intelligence service to provide redacted material, on the topics the political authority specifies, for public release. The service has to comply; it works for the political authority. It can argue for more rather than less bowdlerization of the material for the sake of protecting sources, but it cannot simply say no.

The selective nature of what material gets released in this way is what sways public opinion and influences the outcome of the public debate. The political authority naturally will select whatever supports its case and will leave untouched—and unknown by the public—whatever would undermine the case. However much this may disturb the intelligence service, and however much intelligence officers may believe the chosen policy course to be unwise, they cannot do anything about it. For an intelligence service to release on its own initiative countervailing judgments or information would be seen as insubordination and a move beyond its proper role. The political authority thus gets the benefit of the added cachet that comes from attributing a conclusion to intelligence, while still controlling the substance and direction of the public message.

The very act of redaction can further play to the advantage of the policy maker, by simplifying the message in the desired direction while avoiding doubts and caveats that may be associated with more detailed knowledge of the classified intelligence. That a conclusion is based on intelligence is good enough to retain the cachet that impresses the public. Questions that a consumer of the original classified intelligence might have about the access or reliability of the original source are kept out of the public debate.

These patterns were all exhibited by the public campaign to build support for invading Iraq and overthrowing the regime of Saddam Hussein. Although this campaign came to be associated with faulty intelligence about unconventional weapons, neither this nor any other type of intelligence was the impetus for the decision to launch the war. The war was instead a long-standing project of American neoconservatives, who had been agitating for such an expedition since the mid-1990s. As one prominent neoconservative, Robert Kagan (2005), commented two years into the war:

> I certainly never based my judgment on American intelligence, faulty or otherwise, much less on the intelligence produced by the Bush administration before the war. I don't think anyone else did either. I had formed my impressions during the 1990s.

The election in 2000 of George W. Bush to the US presidency, and the subsequent appointments to senior national security positions in his administration, endowed neoconservatives with the necessary political authority to realize their ambition. Several months later the 9/11 terrorist attack, by suddenly making the American public so militant that it would countenance a measure as drastic as a major offensive war, provided the necessary political climate to do so. The chief values that would be pursued through the war were the core neoconservative values of open politics and free enterprise economics. The main objective of the war to topple Saddam Hussein was not only to establish free politics and free economics in Iraq but also, through regime change in that important Arab country, to stimulate the spread of those freedoms through the rest of the Middle East. President Bush, in a speech three weeks before the invasion of Iraq, stressed that the war would be about the spread of freedom and democracy in the Middle East. He postulated a democratic domino theory, in which 'a new regime in Iraq would serve as a dramatic and inspiring example of freedom for other nations in the region' (Bush 2003).

In addition to political and economic transformation of the Middle East, the political authority's objectives included the exertion of US power as a demonstration of the ability and willingness of the United States to use that power. The idea was to help increase deference to US interests worldwide and to deter adversaries from opposing those interests. While such exertion also flowed from core beliefs of neoconservatism, this objective was probably a major motivation for Vice President Richard Cheney and Secretary of Defense Donald Rumsfeld, both of whom were appropriately categorized less as neoconservatives than as 'assertive nationalists' (Daalder and Lindsay 2003: 15).

As prominent promoters of the war within the US administration later admitted, the issue of so-called weapons of mass destruction was only a convenient topic on which there appeared to be enough consensus to make it the centerpiece of a sales campaign.[1] Before that campaign began, the US intelligence community was barely even raising that subject as a matter of concern for policy makers. The Bush administration never requested the intelligence estimate on this subject that would become infamous (members of Congress from the opposition Democratic Party did so), and President Bush evidently never even read the estimate (Pillar 2011: 35–9). The policy makers were not interested in the intelligence community's judgments as much as they were in using bits and pieces of intelligence reporting that would be most likely to raise concern among the public about Iraq.

The use of the bits and pieces, and the manner in which tussles between intelligence officials and the political authority over public release of intelligence-based information played out, were illustrated by an episode involving a report about supposed Iraqi purchases of uranium ore in Africa. The report was just the sort of tidbit that caught policy makers' attention for possible use in the public sales campaign. The public could understand a simple concept such as buying uranium more easily than it could understand more technical matters such as whether the design of an aluminum tube made it suitable for use as a centrifuge rotor in an enrichment cascade. The White House repeatedly proposed using the report in public speeches. The original source of the report was of questionable credibility, however, and US intelligence analysts had further doubts about the report given that Iraq already had ample supplies of uranium ore. Thus, the intelligence community—in one case in the person of Director of Central Intelligence George Tenet—repeatedly warned the White House not to use the report. Policy makers backed off in the face of those warnings, but the item was so temptingly juicy that they did not give up. They tried again and finally got a lower level intelligence official to accede to public use of the report as long as it was mentioned that Britain, not the United States, had collected it. The report—erroneous, as it turned out—about purchases of uranium ore therefore was inserted into a major presidential address, in a way that treated it as fact and made no mention of the doubts of US analysts (Pillar 2011: 32–4).

The episode demonstrates how the decision to invade Iraq, far from being driven or guided by intelligence, was motivated by other influences, with intelligence only being used selectively and tendentiously to sell the decision to the public. The aftermath of the episode further demonstrated how much the Iraq War story was one of opposition between intelligence and the political authority, notwithstanding how much some intelligence judgments about weapons programs appeared to aid the sales campaign. The Central Intelligence Agency had retained a retired US ambassador to travel to Africa to investigate the allegation about purchases of uranium ore. He concluded that no such purchase had occurred. He was so disturbed by the later use of the erroneous report in a presidential speech that he published an opinion piece in a major newspaper about the subject (Wilson 2003). This was enough of a challenge to the pro-war campaign that the office of the vice president—which was in the forefront of that campaign—began an effort to discredit the ambassador and the agency that had hired him. This led to the blowing of the cover of a CIA officer (who was the ambassador's wife) and the criminal conviction of a senior vice presidential aide for perjury and obstruction of justice.

The power of the political authority to determine what intelligence is made public and what is kept secret applies not only to specific reports but also to entire subject areas. In this respect it is even more apparent how the US decision to launch the Iraq War was made in spite of, not because of, the conclusions of intelligence. The other major theme, besides weapons of mass destruction on which promoters of the war based their public campaign, was the connection of the Iraqi regime to international terrorism. This theme flowed from the terrorist attack of September 11, 2001. Although Iraq had nothing to do with it, the attack was the pivotal event that so drastically inflamed and angered the American public that it made politically possible, for the first time, the launching of a major war of aggression such as the one in Iraq. The theme combined with the one about Iraq's weapons programs to produce the constantly repeated and feared, even if fanciful, scenario about a dictator giving weapons of mass destruction to terrorists. On the issue of relations with terrorism, the differences between intelligence judgments and the public sales campaign went beyond the credibility of individual reports to fundamental judgments. After exhaustively examining (because the policy makers were so keenly interested in the subject) any possible connection between Saddam Hussein's regime and al-Qaeda, the US intelligence community concluded that there was not, as President Bush would assert, an 'alliance' between the two, or anything close to it (Inspector General 2007).

This conclusion did not stop the policy makers, for whom the notion of terrorist connections—especially to al-Qaeda, the perpetrator of the 9/11 attack—was a key part of the pro-war campaign. This is where the distinction between raw intelligence reporting and other information, on one hand, and the judgments of intelligence services, on the other hand, is important. Policy makers and their immediate staffs have access to most of the same raw reporting as intelligence analysts, which means that they are free to stitch selected pieces of that reporting together to convey messages different from the conclusions of the analysts. This is exactly what took place in the run-up to the Iraq War. The center of action in this case was in the Pentagon, and specifically in a small unit of pro-war activists assembled by the civilian policy makers in the Department of Defense.

The function of this unit was not only to construct an alternative to the intelligence community's conclusions about terrorist connections, but also to undermine and discredit the community's work on the subject. The unit put together a briefing, based on selected morsels of reporting with little or no attention to their credibility, in support of the notion that there was extensive operational cooperation between the Iraqi regime and al-Qaeda. The briefing also included a criticism of the intelligence community that purported to explain why the community had missed this supposed alliance. The briefing was presented at the White House and also made its way to Capitol Hill.

In moving beyond the major themes of the public campaign, it becomes even more apparent that intelligence did not influence the decision to go to war in Iraq. What turned out to be much more important than either chimerical weapons of mass destruction or terrorist links were the tribulations that the United States and its allies, not to mention Iraq itself, would suffer for years after the ouster of Saddam. The US intelligence community, on its own initiative, addressed this subject as well. There were three major pre-war assessments about Iraq that received the imprimatur of the entire US intelligence community. The infamous estimate about unconventional weapons was one. The other two concerned the principal challenges that would be encountered in Iraq after Saddam was gone and the regional repercussions of regime change in Iraq (National Intelligence Council 2003a, 2003b). The latter two assessments were gloomy reading. The assessment about challenges inside Iraq, for example, presciently pointed to the high likelihood of violence among Iraq's sectarian and ethnic communities, the low likelihood of stable democracy taking root any time soon, the major expenditures necessary to get the Iraqi economy on its feet, and major flows of refugees. Any objective reading of these latter two assessments could not possibly be deemed as support for going to war in Iraq. To the contrary, they pointed to many of the biggest reasons the war turned out to be a costly blunder.

Those assessments played almost no role in pre-war deliberation inside the government, let alone outside it. Senior policy makers who bothered to glance at them brushed them aside. The assessments obviously were of no help in the mustering of public support for the war, which was the preoccupation at the time. Members of Congress were similarly inattentive, with most of them content to go with the angry, militant post-9/11 flow and to rationalize an offensive war based on an oversimplified idea about weapons of mass destruction.

With intelligence officers powerless to offer any public corrective to this highly unbalanced input to a decision to use armed force, any possibility of correction lay with the legislature. Although in the United States the Congress is a co-equal branch of government that is supposed to share policy making with the executive branch, the politics of the moment did not encourage Congress to play that role effectively with regard to Iraq.[2] The inattentiveness to intelligence extended even to the highly touted issue of unconventional weapons. Only a tiny proportion of members of Congress bothered to look at the intelligence estimate on that subject. One of the few who did—the chairman at the time of the Senate Select Committee on Intelligence—was struck by how much this estimate, filled with dissents and squishy evidence, departed from the Bush administration's public statements; he voted against the Congressional resolution authorizing the war (Graham 2005).

Some members of Congress did pay attention to key parts of the intelligence output and attempted to raise questions that were getting insufficient attention in the march to war, but their efforts ultimately had little effect. One of the questions addressed in the intelligence assessment about Iraqi weapons programs was whether Saddam Hussein, if he did have the feared weapons of mass destruction, was likely to give them to terrorists or otherwise use them against US interests. The judgment of the intelligence community was that he would not—except in the extreme circumstance of his country being invaded and his regime on the verge of being overthrown. Democrats on the Senate Intelligence Committee appropriately seized on this judgment and, given how directly it contradicted the central part of the administration's public pro-war message, insisted that the judgment be made public too. It was, in the form of a letter from the director of central intelligence to the committee chairman that was published in the *New York Times*.[3] But this disclosure did nothing to slow the pro-war campaign. The day the letter was sent to the committee, President Bush made a major speech centered around the scary theme of Saddam Hussein giving the feared weapons to terrorists.

The bully pulpit of the presidency and the ability of the president and his senior lieutenants to inculcate simple themes in the public's consciousness, are far more important in shaping public opinion than conclusions of an intelligence service, even if members of the legislature try to highlight those conclusions. The evolution of public perceptions regarding Iraq and the 9/11 attack clearly demonstrates this. Opinion polls showed that in the immediate aftermath of 9/11 few Americans suspected Iraqi involvement. Several months of the Bush administration's campaign to build support for a war, however, dramatically changed this result. The change did not even require specific assertions by the administration of Iraqi involvement. A rhetorical drumbeat in which 'Iraq', '9/11' and 'war on terror' were repeatedly uttered in the same breath was enough to achieve this result. By August 2002 a majority of Americans thought that Saddam Hussein had been personally involved in the 9/11 attack.[4]

## Policy, intelligence, and the direction of influence

The US initiation of the Iraq War was an extreme and unusual case in some respects, especially in having no policy process inside the Bush administration for determining whether this use of armed force was a good idea. The influence of public statements on public and Congressional opinion was almost the entire story of how the issue of going to war played out. There was no corresponding story inside the administration because there was no opportunity for debate among government agencies and officials over whether launching the war made sense. The only thing to be said about the role in this decision of expert authority—including the intelligence services as well as other sources of expertise inside the government, such as the diplomatic service—is that it had no such role because it never was given a chance to provide input to the decision. The question of whether to launch the war was not on any agenda.

Expert authority outside government also was excluded. Policy makers in the Bush administration brushed aside the offering of help from think tanks and other sources of outside expertise (Pillar 2011: 53–4). Perceptions among policy makers of likely developments in Iraq following Saddam's ouster were developed casually and superficially, based in part on contacts with favored exile groups.

As extreme as the Iraq War case was, it was representative of a larger pattern, at least in the United States, of the non-influence of intelligence on decisions to employ armed force. In looking at such employments over the past several decades, it is hard to discern any significant influence played by intelligence. Of course, this does not speak to situations in which armed force might have been used but was not, although the influences in such situations are inherently more difficult to identify and assess. Possibly, for example, intelligence may have played a role in the hesitancy of the administration of Barack Obama to become militarily involved in the Syrian civil war, although outsiders have no clear way of knowing whether it did or did not.

Among the instances since the Second World War in which the United States did enter an armed conflict, the non-influence of intelligence can be seen in some cases in which there was no reason to expect intelligence, even if it was heard and heeded, to imply that the decision on the use of force ought to come out in a particular way. President Truman's decision to intervene in Korea in June 1950, for example, was arrived at quickly and grounded in lessons he and his advisers drew from the experience in Europe in the 1930s about the need to stand up to aggressors (Truman 1956: 332–3). The more recent decision to intervene in Afghanistan was propelled by the desire, strongly held by Americans generally, to strike back at the perpetrators of the 9/11 terrorist attack and their Taliban hosts. Non-intervention never was a seriously considered option. The subsequent evolution of the expedition in Afghanistan into a long-term nation-building exercise did not reflect a decision that was on the table in 2001.

The large-scale US military intervention in Vietnam in the mid-1960s was a case in which the intelligence implied that non-intervention would have been a better idea. Quite unlike the Iraq War, the Vietnam intervention involved a thorough policy process in which US intelligence had ample opportunity to provide input.[5] A recurring theme of that input was the dim prospect that a stable and strong non-communist South Vietnam would ever be able to establish itself and stand on its own feet. Another theme was the difficulty in ever getting North Vietnam to back down from its effort to unify Vietnam by force. The policy makers' decision to send anyway a force that eventually numbered more than a half million troops partly reflected some disagreement with the intelligence judgments, especially concerning the 'domino theory'—the idea that losing South Vietnam would mean the fall of several other countries to communism. Some of the key policy makers, however, shared a gloomy view about the prospects for ultimate success in South Vietnam; they nonetheless thought it was necessary to intervene because they—and many others—believed that failure even to try would severely damage US credibility.

The superior–subordinate relationship between political authorities and intelligence services that enables the former consistently to trump the latter in major decisions about armed force also raises the issue of the former influencing the work of the latter. This is politicization, not in the sense of the selective public use of intelligence—which was rampant in, but by no means unique to, the case of the Iraq War—but instead in the sense of policy preferences tainting intelligence analysis. This form of politicization gets insufficiently recognized for several reasons, including the disinclination of political leaders to acknowledge that they exerted such influence, the reluctance of intelligence officers to admit that they were so influenced, and a public hope that intelligence failures can be prevented through bureaucratic fixes. The two major official pronouncements on the work of the US intelligence community regarding Iraqi weapons program—one by a Senate committee and the other by a White House appointed commission—refused to recognize politicization. Given the very strong policy preference felt by everyone in the bureaucracy in the lead-up to the Iraq War, however, as well as the multitude of ways in which this climate could and did influence the work of intelligence analysts on the problem, the denial of politicization in this instance is implausible (Pillar 2011: Chapter 6; Rovner 2011). The commission itself uncovered in its interviews direct acknowledgment by intelligence officers that the policy preference affected their judgments.[6] In the United Kingdom, where the policy intelligence dynamic before the Iraq War was similar to what prevailed in the United States, the official findings in the Butler Report, unlike its US counterparts, included an acknowledgment that intelligence judgments and policy preferences had been improperly commingled (Committee of Privy Counsellors 2004).

## Conclusion

The dominance of political over expert authority in influencing decisions to use armed force flows largely from the particular characteristics of this issue area, especially the role of public release of classified information and the politically charged nature of any decision to go to war. Some aspects of this pattern of decision-making, however, are probably shared to varying degrees with other issues. One concerns the role of values and larger objectives, and how they may exert greater weight than facts, assessments, and the other wares of expert authority. The makers of the Iraq War, for example, were motivated largely by a desire to use regime change in Iraq to spread democratic, free market values throughout the Middle East. They were operating according to gut feelings about these values and were not about to be stopped by inconvenient facts and assessments about near-term prospects for Iraq.

There is also the major role played by widely held, largely unchallenged conventional wisdom. Sometimes such conventional wisdom is held by many elites as well as by the general public, and as such it gets reflected in the pronouncements and policies of political authorities. The belief, broadly held at the time of the Vietnam intervention, about what would damage US credibility was perhaps the single biggest influence on the decision to intervene, even though careful examination of history indicates that the belief is largely mistaken. A similar belief is displayed in current debates about the application of military force elsewhere.

## Notes

1 See the remark by one of the most fervent proponents of the war, Deputy Secretary of Defense Paul Wolfowitz, in an interview with *Vanity Fair* in May 9, 2003, transcript at: www.defenselink.mil/transcripts/transcript.aspx?transcriptid=2594
2 A good account of how the US Congress handled the Iraq War issue is in Thomas E. Ricks, *Fiasco: The American Military Adventure in Iraq* (2006), especially pp. 61–4, 85–90.
3 George Tenet, letter to Chairman of the Senate Select Committee on Intelligence, *New York Times*, October 9, 2002.
4 The polling data are cited in Pillar (2003: xviii).
5 The most authoritative account of US decision-making on the Vietnam War is the official study later published as *The Pentagon Papers*, Senator Gravel edition (1971).
6 Commission on the Intelligence Capabilities of the United States Regarding Weapons of Mass Destruction (2005: 75, 107).

## References

Bush, George W. 2003. 'The Future of Iraq.' Speech to the American Enterprise Institute, February 26. Retrieved at: www.presidentialrhetoric.com/speeches/02.26.03.html

Commission on the Intelligence Capabilities of the United States Regarding Weapons of Mass Destruction (US). 2005. 'Report to the President of the United States.' March 31.

Committee of Privy Counsellors (UK). 2004. 'Review of Intelligence on Weapons of Mass Destruction.' Report, July 14. Retrieved at: www.factcheck.org/UploadedFiles/Butler_Report.pdf

Daalder, I.H. and Lindsay, J.M. 2003. *America Unbound: The Bush Revolution in Foreign Policy*. Washington, DC: Brookings Institution Press.

Graham, B. 2005. 'What I Knew Before the Invasion.' *Washington Post*, November 20.

Hessing Cahn, Anne. 1993. 'Team B: The Trillion Dollar Experiment.' *Bulletin of the Atomic Scientists*, 49(3): 24–7.

Inspector General (US). 2007. 'Review of the Pre-Iraqi War Activities of the Office of the Under Secretary of Defense for Policy.' Report no. 07–INTEL–04. Deputy Inspector General for Intelligence, US Department of Defense, February 9. Redacted version can be retrieved at: https://fas.org/irp/agency/dod/ig020907-decl.pdf

Kagan, R. 2005. 'On Iraq, Short Memories.' *Washington Post*, November 20.

National Intelligence Council (US). 2003a. 'Principal Challenges in Post-Saddam Iraq.' Washington, DC: National Intelligence Council. Redacted version can be retrieved at: http://intelligence.senate.gov/11076.pdf

National Intelligence Council (US). 2003b. 'Regional Consequences of Regime Change in Iraq.' Washington, DC: National Intelligence Council. Redacted version can be retrieved at: http://intelligence.senate.gov/11076.pdf

Office of the Director of National Intelligence (US). 2007. 'Iran: Nuclear Intentions and Capabilities.' November.

*Pentagon Papers*. 1971. Senator Gravel edition, 4 vols. Boston, MA: Beacon Press.

Pillar, P.R. 2003. *Terrorism and U.S. Foreign Policy*. Washington, DC: Brookings Institution Press.

Pillar, P.R. 2011. *Intelligence and U.S. Foreign Policy: Iraq, 9/11, and Misguided Reform*. New York: Columbia University Press.

Ricks, T.E. 2006. *Fiasco: The American Military Adventure in Iraq*. New York: Penguin.

Rovner, J. 2011. *Fixing the Facts: National Security and the Politics of Intelligence*. Ithaca, NY: Cornell University Press.

Russett, B. 1994. *Grasping the Democratic Peace: Principles for a Post-Cold War World*. Princeton, NJ: Princeton University Press.

Truman, H.S. 1956. *Memoirs, vol. 2: Years of Trial and Hope*. Garden City, NY: Doubleday.

Wilson, J.C. 2003. 'What I Didn't Find in Africa.' *New York Times*, July 6. Retrieved at: www.nytimes.com/2003/07/06/opinion/what-i-didn-t-find-in-africa.html

# 4 The long goodbye

## Science and policy making in the International Whaling Commission

*Michael Heazle*

Founded on the 1946 International Convention for the Regulation of Whaling (ICRW), the International Whaling Commission (IWC) is both the longest running global wildlife management regime and the first to explicitly acknowledge scientific authority as the preeminent source of policy legitimacy. Article V of the Convention states 'amendments of the schedule [i.e. IWC policy] . . . shall be based on scientific findings' (IWC First Report 1950: 11). But contrary to the importance assigned to science by parties to the convention in 1946—a time when support among whaling states for Antarctic whaling to be *sustainably* managed appeared strong and unequivocal[1]—the science-based management advice offered by the commission's Scientific Committee seldom has exerted the level of policy influence envisaged by the IWC's founding states. Rather than help to rationalize policy making by 'speaking truth to power',[2] the advice and expertise of the Scientific Committee instead has been very much 'on tap' and under pressure to be 'in line' with the national and domestic political interests of the commission's most influential states.

Indeed, the political authority generated by strong domestic political preferences and national interest perceptions within both whaling and anti-whaling liberal democratic states (e.g. whales as an important consumer and/or cultural resource versus 'the whale' as an important environmental symbol),[3] have caused the credibility and influence of expert authority-based policy advice to rise and fall within the commission over time. Moreover, and as I argue in this chapter, the IWC serves as a powerful illustration of how expert authority (EA) can become almost entirely irrelevant as a restraining influence on political interests and imperatives in policy decision-making when levels of uncertainty in expert advice are high and both policy ends *and* means become the sites of entrenched and ongoing political conflict.

The rapid depletion of great whale stocks in the Antarctic during the 1950s (blue, humpback, fin, and finally sei whales) briefly re-established the authority of science in the commission during the 1960s after almost a decade of near irrelevance in the face of growing uncertainty over stock numbers and industry demands for high catch quotas. But the Scientific Committee's new-found influence was to be short lived. From the early 1970s onwards—when perceptions of whales and whaling were undergoing fundamental change among many in the

IWC's rapidly expanding membership—uncertainty arguments were being made to support a moratorium on *all* commercial hunting, despite the many improvements in management methods achieved by the Scientific Committee and its strong opposition to a blanket ban approach to whaling (Heazle 2006, 2010; Aron 2001a; Bannister 2001; Schweder 1992). Thus, the influence of anti-whaling politics in most Western states, particularly those with no whaling interests, quickly replaced industry interests to become a pervasive influence on policy debate in the IWC. Indeed, the IWC's major management debates to date—the struggle for lower catch limits in the 1950s and 1960s; the adoption of the moratorium in 1982; the commission's reluctant acceptance in 1994 of a new management regime, the Revised Management Procedure (RMP); and the ongoing debates on and subsequent delays to the RMP's implementation—all have been driven mostly by arguments citing different perspectives on whales and whaling, depending on the interests and domestic political imperatives of the actors concerned, rather than advice from the Scientific Committee aimed at achieving sustainable catch quotas for both whales and whaling interests.

The IWC's divided and ultimately deadlocked policy environment, in which even the commission's fundamental purpose and goals are no longer commonly understood, can, I further contend, be understood as an illustration of Martin Hajer's (2003) description of how an 'institutional void' can occur when traditional principles and expectations of legitimate policy making are overwhelmed by highly complex, multi-framed, and broadly contested policy dilemmas, in particular those institutional principles and practices relying on the ability of expertise to rationalize and help resolve political conflict over ends and means. The IWC's institutional void is, I further argue, characterized by an ongoing series of 'legitimacy crises' that can be linked directly to the political and expert authority tensions created by not only the conflicted nature of the ICRW's formal, institutional standards of legitimate policy (i.e. policy informed by *both* scientific advice *and* industry interests), but also by the unreasonably high expectations placed on expertise by the convention as an effective arbiter of conflict between those standards. But while Hajer's explanation of how and why institutional voids occur is underwritten by an expectation that institutions will evolve through the negotiation of new means and principles for managing the new challenges and 'political spaces' thrown up by contemporary society, the IWC experience with attempting to manage whaling suggests that institutions may instead become paralyzed if political conflict becomes too entrenched, raising the possibility of institutional failure rather than evolution in the face of mounting uncertainty and myriad political framings and interests.

The following section examines the estrangement of expert authority in the IWC, in spite of the strong role given to it by the commission's founding convention, within Hajer's framework, highlighting the IWC as an example of how high and sustained levels of political conflict can prolong policy deadlock within an institutional void over several decades, leading ultimately, perhaps, to institutional failure and collapse. The remainder of the chapter then traces the major debates and policy decisions within the IWC over the last 60 years to illustrate 1) the kinds of institutional challenges the IWC institutional arrangements have struggled to manage; and 2) how recognition among IWC member governments of the Scientific Committee's expert authority as *the* legitimate basis for policy has been restricted to rare junctures in the commission's history where political consensus among members over management priorities and goals occurred. The one brief period where the authority of scientific expertise was strong enough to influence commission policy—the late 1960s and early 1970s—occurred when neither strong political opposition to its advice nor division over management priorities existed.

## The IWC's institutional void?

Hajer's (2003, 2009) concept of an 'institutional void' refers to situations where, according to Hajer, classical–modernist-based institutional arrangements, in particular those relying heavily on the authority of experts as a key provider of policy legitimacy, fail to manage the complex knowledge and political problems that often arise from new types of policy issues and challenges.

> In an inherently dynamic society the legitimacy and efficacy of the particular set of institutional arrangements [democratic and expert authority, ministerial responsibility, role of bureaucracy and constitutional rules] might be challenged by new developments . . . Indeed in the light of new problems and new problem perceptions, these institutions might simply lack the authority or focus needed for problem solving that is widely perceived to be both effective and legitimate.
>
> (Hajer 2003: 177)

In the IWC—an international government organization whose membership reflects the values of many different societies—an institutional void has emerged out of a fundamental tension between the ICRW's formalized institutional expectations of science and the policy objectives it would serve, and the limits imposed on expert authority by the complexity and inherently political nature of the issue area the ICRW requires science to successfully manage. This tension has been most clearly demonstrated by the competing, and often incompatible, priorities of the commission's membership and their resistance to any negotiated compromise or resort to expert authority. In this sense, the IWC serves as an important illustration of not only how unrealistic expectations of expertise in complex and contested policy areas can undermine rather than promote legitimacy in policy, but also how difficult it can be to alter existing institutional arrangements concerning the role of science and expertise in policy making once political divisions become entrenched to the point where they remain resistant to negotiation.

Indeed, the IWC's institutional rules and objectives, and their heavy formal reliance on EA as the main source of policy legitimacy, have remained both ineffective and unchanged over a very long period of time. Moreover, the commission's institutional arrangements appear unable to be altered in ways that may allow it to better cope with the elements Hajer (2003: 177–80) has identified with the emergence of an institutional void, in particular the undermining of 'classical expertise' through modernity's expansion of uncertainty; the larger and more participatory scope for policy involvement by various actors that accompanies the growth of civil society (both domestically and internationally); and the more expansive and interconnected range of issues that policy making must deal with as interaction between actors and issue areas alike becomes more varied and diffused. Environmental issues like whaling and climate change, along with biotechnology and genetics, are core examples of the kinds of fast-evolving and largely unprecedented policy problems that most challenge the Enlightenment informed rules and principles identified by Hajer (2003: 176–7) as characterizing most modern political institutions. Hajer's point, put simply, is that the evolution of those institutions was contingent on the assumptions and policy priorities of a time when perceptions of policy issues and expectations for legitimately dealing with them were very different. As a consequence of this historical contingency, existing institutions become increasingly exposed to policy making challenges they are ill-suited to manage as the ever-expanding impacts of science and technology on both society and individuals throw up ever-growing levels of uncertainty and complexity; classical–modernist institutions are, as it were, the creation of a different time and, therefore, need to evolve if they are to remain relevant.

On this point Hajer suggests that institutional voids also should lead to the creation of new rules and principles in the course of negotiating responses to new problems, what he describes as 'new emergent practices of governance' (Hajer 2003: 189). Thus, presumably new, if not only reformed, institutions and arrangements develop over time that are better suited to managing contemporary policy challenges. But, as the IWC experience over more than 60 years of internal division and policy conflict indicates, institutional voids, once in place, may instead result in only an ongoing series of legitimacy crises if a balance between expert and political influence on policy making, and agreement on how they are understood, cannot be negotiated as the basis for new rules of the game being developed. The IWC's inability to reconcile its institutional framework and the sources of policy legitimacy it demands with first the scientific complexity of whaling management, and then later the political complexity brought about by changing perceptions, multiple actors, and domestic political imperatives, has created a *legitimacy vacuum* within the commission that has left it deadlocked and unable to manage whaling in any meaningful sense.[4] Instead of reaching agreement on institutional renewal and reform, IWC policy makers have remained locked within the IWC's existing institutional arrangements by seeking legitimacy for their positions through the 'scientization' of policy debate (Sarewitz 2004)—that is, attempting to hide or obscure the political drivers of policy disagreement behind debates over competing scientific claims in order to still appear consistent with the institutional requirements of the ICRW.

The work and recommendations of the Scientific Committee are necessarily aimed at promoting sustainable hunting of whales due to the purpose and obligations of the commission established by the ICRW,[5] and the convention has continued to require commission policies to be based on those findings. But in practice most governments at one time or another have instead sought to use the Scientific Committee—and the policy authority and legitimacy given to it by the convention—as a vehicle for reconciling their formal commitment to EA derived legitimacy under the ICRW with the more electorally significant political authority (PA) derived legitimacy expectations of domestic voters and interest groups. According to long-serving IWC scientist Doug Butterworth (1992: 532):

> The real debate in the IWC has been between some countries wishing to preserve industries, employment and a food source based on whales, and others wanting these animals classed as sacrosanct. The terms of the convention [the ICRW] have required that this debate be conducted in a scientific guise, so that these hidden agendas have had to be played out in the scientific committee.

In the IWC, tensions between EA and PA were demonstrated most clearly by the overhunting of the 1950s and early 1960s, and then later by the like-minded anti-whaling majority within the commission that formed in support of the moratorium proposal during the late 1970s and early 1980s. These examples show why the expectations placed on the Scientific Committee by the ICRW—that is, to act as a neutral magistrate in a complex market place of conflicting values, interests, and ideas—are ill-conceived given not only the many uncertainty-based challenges to the Scientific Committee's authority to influence policy made by industry interests during the 1950s, but also the more normative challenges later raised by anti-whaling advocates that led to the very notion of whaling itself becoming a deeply contested issue. By the late 1980s, whaling, in the traditional sense of the term, had become so politically conflicted that an influential minority of members were arguing that sustainable management meant *no* whaling—a view encouraged by both the decline of whale products as an important economic resource and the altered political status of whales as powerful environmental symbols and endangered mega-fauna.

The IWC's institutional arrangements initially required a level of authority to guide policy that could not be realized given both the weakness of the available science in the 1950s and the huge financial pressures driving industry interests. But even when the expertise provided by the Scientific Committee reached a point where it could more confidently respond to the questions and demands produced by debate in the commission over sustainable catches in the 1970s, its authority again was challenged, this time by the willingness of particular members or groups within the commission to use either uncertainty or moral arguments, or both, to undermine the credibility of the expert advice on offer, depending on its fit with the political priorities being pursued.

By 1993, when a majority of members rejected the Scientific Committee's unanimous advice to adopt the RMP, a growing number of influential IWC governments were now opposing any return to commercial whaling—regardless of the scientific advice on offer—and were increasingly being perceived by pro-whaling governments and groups to be making demands for excessive levels of precaution. This division over goals and acceptable levels of risk has persisted, in spite of the commission's adoption of the RMP, and now prevents agreement on the Revised Management Scheme (RMS), which provides the inspection and verification rules under which the RMP must operate. The continuing dearth of compromise over the level of precaution the RMS rules should provide for (e.g. international observers on all or only some whaling vessels; DNA testing of whale meat) has led to hunting continuing, both commercially and for research purposes, outside the international management controls the IWC is supposed to provide—thereby making the IWC largely redundant as a wildlife management organization. Furthermore, the ongoing impasse may well result in the whaling nations eventually leaving the commission to whale under regional organizations—a move that effectively would eliminate any further prospect for global regulation of any future commercial whaling.

## Management and science in the 1950s: great expectations betrayed

The as yet unresolved tension between the leading role given to science by the ICRW as the key source of policy legitimacy and its *actual* role in commission policy making was written, perhaps unwittingly, into the convention by its founding members. An additional provision in Article V requires that policy decisions, in addition to being based on 'scientific findings', also take into account *the interests of whale product consumers and the whaling industry* (IWC First Report 1950: 11). The ICRW thus requires a balance to be struck between sustainable exploitation and industry interests, but in doing so gives no indication as to how such a balance should be achieved other than through the advice of experts.

The initial outcome of this ambiguity was that the IWC's first decade of attempting to sustainably manage Antarctic whaling became, by any standard, a disaster. Scientific research in the IWC, in spite of the important role given to it by the ICRW, mostly remained on the fringes of the commission's policy making until the late 1950s when the threat of whaling's imminent collapse finally became sufficiently convincing—due to the increasingly poor catches by the fleets—to prompt at least a partial change in the commission's perception of uncertainty and the role of scientific advice in commission policy. The relatively minor role afforded to scientific advice in practice was mostly due to the IWC's strong pro-industry bias and the weakness of cetacean science at the time, which prevented it from countering the dominance of industry interests among the governments represented at the commission.

In the 1950s, for example, the five Antarctic nations (the UK, Norway, Japan, the Netherlands, and the USSR) all, to varying degrees, based their opposition to quota reductions on both the impact they would have on their respective whaling industries and the many uncertainties surrounding the Scientific Committee's claim that stocks were being overhunted. However, the Scientific Committee's warnings were soon vindicated by the near total collapse of the Antarctic stocks and also whaling as a viable industry for all but two of the Antarctic nations (Japan and the USSR).[6]

Thus, in the first decade of IWC stewardship, it quickly became clear that the available scientific knowledge and understanding of whale species and their reproduction capacities under hunting was too limited for the Scientific Committee to play the role expected of it. A combination of numerous and often irreducible management uncertainties and opposition from governments to quota reductions faced by the commission's scientists effectively undermined their authority to shape commission policy as the ICRW had intended, rendering the Scientific Committee incapable of reining in either the increasing numbers of whales taken, or the growing levels of investment that drove the overhunting. Rather than 'speaking truth to power', the Scientific Committee was instead reduced to mostly unsuccessful attempts at defending its advice *from* power, and sometimes even adapting this advice to the political interests of governments in order to remain relevant.

Indeed, the IWC scientific community was small and somewhat isolated during the 1950s and while many of its members were competent in cetacean biology, few had any expertise in the still developing fields of statistical analysis and population dynamics. As Schweder (1992) has noted, the post-war period saw significant progress in statistical methodologies and fisheries science, culminating in Raymond Beverton and Sidney Holt's landmark 1957 work, *On the Dynamics of Exploited Fish Populations*. These fields have since provided the basis for the development of cetacean management science (Schweder 1992: 6). The IWC's Scientific Committee, however, was seriously restricted in its ability to take advantage of such expert advice by budgetary pressures that, in addition to thwarting attempts to include members of the broader scientific community in its deliberations,[7] forced its own members to meet 'somewhat irregularly' during the 1950s (Allen 1980: 26) and also delayed publication of the Scientific Committee's reports to the commission until 1955.

During the 1950s, the Scientific Committee's deliberations ostensibly were focused only on the biological issues pertaining to management as described by Article IV of the convention (IWC First Report 1950: 11), but several instances during the 1950s demonstrated that the Scientific Committee's advice on biological issues was often tempered by non-scientific factors. At the IWC's sixth meeting in 1954, for example, a Scientific Sub-Committee proposal for separate, species by species quotas rather than the non-discriminating total catch based on the Blue Whale Unit (BWU)[8] was supported on the grounds that 'it would be a great advantage' in terms of conservation. But the sub-committee agreed not to put the proposal forward after a Norwegian member reminded his colleagues that 'there would be great practical difficulties in operating such separate quotas' (IWC Sixth Report 1955: 18–19). Then, in 1955, alarmed by the rapid decline in blue and fin stocks, the Scientific Committee believed that, in addition to separate quotas, a drastic cut in the quota from 15,500 to 11,000 BWU was needed. But as in the previous year, the scientists realized that the commission would not accept such a sudden reduction due to the hardship it would cause the whaling companies. Instead, they recommended reducing the quota incrementally, starting with 14,500, but even this watered down proposal was rejected by the commission (see IWC Seventh Report 1956: 23–4; Birnie 1985: 227–8).

The question of why the IWC's scientists allowed their advice to be so heavily influenced by what they thought the commissioners would or would not accept is an important one—particularly in terms of the implications it has for ideas about the proper role of science in resource management—and can be answered in part by the lack of population dynamics expertise during the 1950s. But the main obstacle to the Scientific Committee gaining greater authority over policy during the 1950s was the widely shared view within the IWC, and by some scientists within the Scientific Committee, as illustrated by the sustained opposition of the Dutch scientists to quota reductions,[9] that the commission's main priority was to provide the industry with maximum financial return in the shortest time possible. The political imperatives for doing so were made clear by the importance of whale products during the initial post-war period, both as a food and industrial resource, and also the huge sums of money invested in whaling by governments and private interests (Tønnessen and Johnsen 1982).

Some observers believe that the Scientific Committee should have been more forceful, and lay much of the blame for the excesses of the 1950s at the feet of the commission scientists. Cushing (1988: 162), for example, writes:

> the ultimate blame [for excessive catches] must lie with the scientists, who ignored what had happened in the past. Perhaps they were overwhelmed by their lack of exact information and lacked the will to give good advice without it.

However, Cushing's assessment is unduly harsh. Moreover, it betrays an unjustified faith in the authority of experts to control the willingness of governments to pursue key interests. The flawed institutional arrangements of the IWC had pitted a newly emerging and highly uncertain area of scientific knowledge against powerful industry interests, leaving the authority of the commission's scientists highly vulnerable to the many questions raised by the Antarctic whaling states, which they could not confidently answer. Indeed, given the constraints the Scientific Committee was facing, and also a historical record that shows the majority of scientists did attempt to warn the commission of the extent to which they *believed* the Antarctic stocks were being depleted, it is hard to imagine any other outcome. Had the commissioners been prepared to listen to the majority of scientific opinion concerning the status of the blue, fin, and humpback stocks, a more conservative approach to the setting of quotas no doubt could have been taken. The majority of member governments, however, clearly were not prepared to take on any advice that would endanger their industries, and it is for this reason that a more forceful stand by the scientists was unlikely to have made much difference, even if they had been able to speak with greater authority. On several occasions, the Scientific Committee members, with the notable exception of the Dutch scientists (who regularly opposed calls for lower catches on uncertainty grounds), made their belief in the need for much smaller quotas very clear. But the growing competition and financial pressures among the Antarctic nations meant there was little chance their advice would be heeded, even though a sharp decline in catching among the Antarctic fleets had by this time become obvious. The Antarctic nations in effect had put themselves in a situation where they simply could not afford the kind of scientific advice that was being offered, so they rejected it, regardless of the available evidence.

## A (briefly) united commission turns to science: 1960–1972

By the late 1950s the commission was in disarray following a series of protracted disputes between the five Antarctic nations brought on by a sharp decline in Antarctic catches. The disagreements, which threatened to take whaling permanently out of the IWC's control, concerned the size of relative shares within the Antarctic quota under a new national quotas scheme that earlier had been proposed by Norway to reduce competition between the fleets, and also the need for a reduction of the total quota being argued for by the Scientific Committee. In an attempt to resolve the impasse and preserve the joint influence of all five nations over the setting of quotas, the UK commissioner resorted to invoking the authority of experts. At the 1959 meeting he proposed that

> a small committee of three scientists qualified in population dynamics or some other appropriate science should be appointed by the Commission to carry out an independent scientific assessment of the condition of the whale stocks in the Antarctic which would provide a scientific basis for the consideration of appropriate conservation measures by the commission.
>
> (IWC Twelfth Report 1961: 16)

The UK government's hope was that the credibility of external experts would be enough to establish a political consensus on the size of annual quotas based on scientific advice that everyone, in particular the Dutch, would accept as valid. The UK's resort to science in order to resolve the disputes was, however, ironic since each of the Antarctic nations so far had supported keeping the BWU quota at a limit that satisfied the needs of their industries, despite numerous warnings that the Antarctic stocks were being over hunted. Now, for the first time in the entire history of whaling, catching on the basis of what was believed to be available, rather than simply what was needed, was being given priority on the assumption that an independent assessment by recognized specialists could provide authoritative and compelling estimates of population and sustainable yield for the great whales.

On the basis of this assumption at the 1960 meeting the commission adopted not only the UK proposal for the appointment of what would become known as the Committee of Three (COT, later the Committee of Four), but also the provision that 'the Antarctic catch limit should be brought into line with the scientific findings not later than 31st of July, 1964, having regard to the provisions of Article V (2) of the Convention' (IWC Twelfth Report 1961: 16).

An *Ad Hoc* Scientific Committee (AHSC), formed at the 1959 meeting, was also given an important role in the commission's attempt to establish science-based, sustainable quotas. This committee met for the first time in May 1960, one month before the plenary of the twelfth meeting in London where, because of the absence of the Dutch representative (due to illness), the AHSC was able to unanimously endorse a paper by UK scientist Laws which concluded—on the basis of declining catch per catcher-day's work figures and increasing numbers of younger and immature fin whales in the catches—that fin whale numbers were indeed dropping (IWC Eleventh Report 1960: 23). After concurring that the 'combined evidence leaves no room for doubt of a decline of the fin whale stock in the Antarctic', the AHSC members went on to warn that they were 'convinced that the stock cannot withstand the present rate of catching and that the only remedy . . . is a substantial reduction in the total catch together with continued biological examination of the condition of the stock from year to year' (IWC Eleventh Report 1960: 25–6).

In order to further verify the now 'unanimous' opinion of the IWC's scientists that current exploitation levels were unsustainable (IWC Twelfth Report 1961: 14), the commission, on the advice of the Scientific Committee, instructed the AHSC to work in conjunction with the COT scientists and 'carry out a detailed and specified programme to improve the collection and interpretation of data including the use of the latest methods of studying animal populations' (IWC Twelfth Report 1961: 16). Thus, it was the commission's intention to have the AHSC, with the assistance of the COT, apply the latest available methods in the newly emerging field of population dynamics to a larger and better organized body of data. In addition to the task of assisting the AHSC, the three scientists were asked to 'report within one year of their appointment the sustainable yield of [Antarctic stocks] in the light of the evidence available and on any conservation measures that would increase this sustainable yield' (IWC Twelfth Report 1961: 16).

When the IWC commissioners gathered for the 1962 meeting in London, they were presented with a strong and unequivocal vindication of the past warnings made by the majority in the Scientific Committee and its sub-committees, and also a new and more urgent warning that the IWC's management approach needed to change drastically. As had previously been the case in the 1950s, no one was able to 'prove' that the Antarctic stocks were in decline; the difference, however, was that the empirical evidence of over-hunting (i.e. a sharp and sustained drop in catches) had reached the point where no one in the Scientific Committee was able, or perhaps willing, to offer an alternative explanation of what it meant. The catch per unit of effort (CPUE) data now indicated a clear drop in catches, but only after the damage, it seemed, had already been done; the challenge facing the IWC was not only to prevent the current situation deteriorating further, but to also prevent it from being repeated in the future.

The COT's Second Interim Report's findings, which confirmed those of the AHSC's earlier meeting in 1962 and then were themselves reaffirmed and explained in more detail in the committee's Final Report, warned that 'sustainable yields of blues and humpbacks of any appreciable magnitude could only be obtained by complete cessation of catching of these species for a considerable number of years'. Furthermore, it supported fears that the fin stocks—upon which Antarctic whaling now heavily relied—were also depleted, stating that in order to avoid 'further decline of the fin whale stock and allow it to build up to levels which will sustain high yields, the catches in the next few years must be drastically reduced to something less than 9,000 [i.e. 9,000 animals or 4,500 BWU]' (IWC Fourteenth Report 1964: 38).

The COT's Final Report and Supplementary Report the following year, supported by the Scientific Committee Report, presented even worse news for the IWC—news that was unlikely to have surprised anyone in the commission by this stage. After having assembled the most comprehensive compilation of catch and biological data as the basis for the most detailed and best quantified estimates of stocks and recommendations for sustainable catch quotas ever presented by any group of scientists to the IWC, the COT, in their Final Report, informed the commission that, in the unanimous opinion of the committee, there was no longer any justifiable doubt (or at least none that anyone was prepared to pursue) of the 'serious danger of extermination' currently facing blue whales and some humpback populations, or of the fact that fin whales were 'far below the levels of maximum sustainable yield' (IWC Fourteenth Report 1964: 40).

The COT scientists also repeated the recommendation they had made in their Second Interim Report for the elimination of the BWU and the establishment of separate quotas for each species. Indeed, the IWC's ongoing preference for the BWU, despite calls as early as 1955 by the Scientific Sub-Committee for quotas to be managed on a species by species basis (IWC Sixth Report 1955: 18–19), was a good illustration of the commission's simplistic approach to management, and also its generally low regard for the concept of maximum sustainable yield (MSY) and the long-term benefits it might offer in terms of stable, sustainable catch quotas. At this juncture, with the main Antarctic stocks generally accepted to be at dangerous levels, the need for more selective regulation of catching was obvious, and both the COT's Second Interim Report and Final Report made this point very clearly to the commission (IWC Fourteenth Report 1964: 37, 41). More importantly, the evidence was at last sufficiently convincing to compel a more conservative management approach. Some of the credit for this sea change in the commission's priorities clearly belongs to the work (and tenacity) of the Scientific Committee and the COT. But the biggest driver of change, for the three European Antarctic nations at least, undoubtedly was the realization that there were no longer enough whales left to make whaling profitable.

From the early 1960s onwards, when the number of active whaling states began to fall, the economic and financial considerations that so far had driven IWC policy gradually became less and less relevant to an increasing number of the commission's members. During the latter half of the 1960s only Soviet and Japanese whaling in the Antarctic continued, following the withdrawal of the Dutch (the Netherlands, following New Zealand's lead a year earlier, even went so far as to resign from the IWC), British, and finally the Norwegians from Antarctic pelagic whaling (see IWC Twentieth Report 1970: 10; IWC Twenty-First Report 1971: 11). Thus, the declining relevance of industry interests among the IWC's members—with the exceptions of Japan and the Soviet Union's ongoing operations in the Antarctic and the North Pacific—led to a proportionately increasing level of commitment in the commission to the views of the IWC scientists who for some time had been calling for reduced catches. The rapidly declining catches of the late 1950s and early 1960s combined with the absence of any pressing domestic or economic reasons to continue hunting, in effect, had encouraged the belief that science aimed at conserving stocks was now very relevant if whales and whaling were to survive. By the late 1960s the importance of whaling for many consumers already had ended with whale oil's replacement in the production of margarine by vegetable oils during the 1960s, and also the replacement with synthetic substitutes of baleen and sperm oil in other products.

The decline in both the commercial importance of whaling products during the 1960s and Antarctic whale numbers during the 1950s led to the UK, the Netherlands, and Norway withdrawing from Antarctic whaling; by 1966 only Japanese and Soviet fleets remained. Together with whaling's lack of cultural relevance among the IWC's predominately Western membership, the drop in both consumer demand and whale numbers essentially meant that research arguing in favor of smaller and better regulated hunts was no longer in conflict with the political and economic interests of most member states. The declining relevance of whaling, therefore, set the stage for a short period in which scientific advice was given centre stage in the IWC, and significant progress in research and management methods was made. Indeed, the developments of the 1960s, combined with the emergence of a new awareness of environmental issues in the early 1970s, were redefining the priorities of a growing number of IWC members—a trend that would continue to alter the IWC's political environment for some time to come.

## The Scientific Committee's drift towards redundancy: 1972–1982

The term 'commercial whaling' had not been used in the IWC until copies of a communication from the 1972 United Nations Conference on the Human Environment, held in Stockholm, calling for 'a ten year moratorium on commercial whaling' were distributed at the IWC meeting in London later that year (IWC Twenty-Fourth Report 1974: 24). The London meeting featured an address from Maurice Strong, the Secretary General of the UN conference. Strong outlined to the plenary the conference's resolutions on cetacean management, informing the Commission that the Stockholm meeting had recommended 'the strengthening of the International Whaling Commission, increasing international research efforts, and . . . an agreement involving all governments concerned in a ten-year moratorium on commercial whaling' (IWC Twenty-Fourth Report 1974: 23–4). The US delegation responded by proposing, with support from the UK, 'a motion that the schedule for 1973 be amended in every case where a numerical quota appears to substitute the numeral "0" for all such numerical quotas' (IWC Twenty-Fourth Report 1974: 24).

The US government's push for a *temporary* end to all commercial whaling was among several steps it had recently taken to respond to domestic environmental concerns and, in particular, growing calls to provide protection for the great whales, including several unilaterally imposed regulations passed by Congress that further reflected the changing political mood towards environmental issues in the US and elsewhere (Epstein 2006). As former IWC Commissioner, Scientific Committee member, and Director of the Office of Ecology and Environmental Conservation at the National Oceanic and Atmospheric Administration William Aron (2001b) noted:

> The moratorium appears almost out of nowhere in the early seventies, but one has to put it into a much broader perspective . . . In the Nixon administration—almost at one time—we passed the National Environmental Policy Act; we created the Environmental Protection Agency; we created the National Oceanic and Atmospheric Administration; we changed our water and air quality laws—all for the benefit of the environment . . . During that period, there was—at least in the United States—a wide variety of environmental organizations, many you might expect [were] contrary in interest to one another's views. . . . At the same time there was a strong desire to build an environmental ethic. And the whaling issue was ready made for pulling these organizations together.

The US position took the COT's earlier urgings for the stocks to be given the benefit of any doubts to its management extreme. And although the moratorium proposal was defeated in both the Technical Committee and later the plenary (four delegations voted in favor, six against, and four abstained), its introduction and minority support were enough to set the stage for a debate that soon would dramatically alter the IWC's management priorities and also fundamentally question the Scientific Committee's ability, and therefore also its authority, to manage whaling. At the 1972 IWC meeting, the US delegation, when proposing the adoption of zero quotas in the Technical Committee, argued that 'the state of knowledge of the whale stocks was so inadequate that it was only common prudence to suspend whaling; this was necessary so that scientific efforts could be redoubled and new research techniques developed' (IWC Twenty-Fourth Report 1974: 24). The purpose of the moratorium, therefore, as argued by its supporters, was not to permanently end 'commercial' whaling, but only to pause hunting until it could be made safer by better science.

Already existing divisions within the Scientific Committee over the status of the Antarctic stocks, in particular the fin whale, were no closer to being resolved at the 1972 meeting with the Japanese and Soviet scientists continuing to support catches that were larger than what others in the Scientific Committee were prepared to accept.[10] Sufficiently reliable data for more accurate and convincing estimates, however, remained out of the Scientific Committee's reach. In addition to the ongoing disputes over the status of the Antarctic stocks, fears that fin, sei, and more recently sperm whale stocks in the North Pacific and elsewhere were coming under threat were also increasing, adding further weight to a growing body of opinion outside the IWC that US and UK concerns for the future of the great whales were probably justified. Indeed, the UN sponsored demand for a temporary end to commercial whaling clearly had demonstrated the extent to which the IWC's policies were now attracting international attention (Bailey 2008), causing a dramatic shift in the commission's balance of voting power.

Up to the early 1960s all IWC members represented active whaling countries, with the exception of Mexico. However, in the decade preceding the Stockholm conference, the number of active whaling countries steadily declined until by 1972 only eight of the IWC's 14 members maintained commercial whaling industries. By the early 1980s, the commission was made up of only 10 whaling members compared to 30 non-whaling members (see Gambell 1999: 184). Thus, in lieu of the IWC implementing a management program broadly accepted as reliable and safe, it appears to have been inevitable, in retrospect at least, that political pressure for the moratorium's adoption—both within and outside the commission—would continue to grow.

It is also ironic, however, that calls for a moratorium on commercial whaling began at almost precisely the same time that the Scientific Committee had finally succeeded in having the IWC abandon the now notorious BWU and adopt species by species quotas. At the 1972 meeting provision for the BWU quota was deleted from the schedule and all quotas were subsequently set on a species by species basis for Antarctic baleen whale stocks, a move that allowed MSY now to become the commission's guiding management principle in both theory and practice. Given the substantial improvements in the IWC's approach to management that had occurred by 1972, in particular the abandonment of the BWU system, the joint US/UK push for a blanket moratorium covering all stocks in all regions was not well received by the Scientific Committee, which had long been advocating species by species management (Bannister 2001). In response, the Scientific Committee rejected the moratorium and called instead for a program of more intensive research that would later become known as the International Decade of Cetacean Research (IWC Twenty-Third Report 1973: 38).

As the political momentum behind the policy pendulum's initial shift away from allowing some hunting continued to gather force, it soon was on its way toward the alternate extreme: a global moratorium on hunting. At the IWC's

1973 meeting, again held in London, the US delegation picked up where it had left off at the previous year's meeting by once more proposing the adoption of a ten-year moratorium on commercial whaling. Again, the US proposal failed to gain the necessary three-quarters majority support with the Scientific Committee also repeating its objections to the moratorium from the previous year. What was remarkable, however, was that for the first time in IWC history a simple majority of members—eight in favor, five against with one abstention—had voted in the plenary in support of a global ban on all commercial whaling (IWC Twenty-Fifth Report 1975: 26).

The management priorities of many in the IWC clearly were undergoing dramatic change, and already had altered markedly among several members in the short space of only a year. Furthermore, not only was the US government reinforcing its opposition to the commercial interests of the relatively few IWC members still committed to whaling, pelagic or otherwise, it now was attempting to do so by directly questioning the Scientific Committee's ability to advise on management proposals. After again basing the need for a moratorium on the belief that 'knowledge was inadequate to protect the species'—a point supported in part by four US scientists now advocating protection of Antarctic fin whales on similar grounds (see Fox *et al.* in IWC Twenty-Fourth Report 1974: 190–93)—the US Commissioner went on to state that his delegation also 'had reservations in respect of the recommendations of the Scientific Committee' (IWC Twenty-Fifth Report 1975: 26). The short period of relative calm and general willingness to adhere to the Scientific Committee's advice that had existed in the IWC since the late 1960s was, it seemed, almost at an end.

The IWC's implementation of the New Management Procedure (NMP) in 1975, which Australia had proposed at the 1974 meeting,[11] provided some hope among IWC scientists that the Scientific Committee might yet perform the role accorded to it in the ICRW. The US and UK had agreed to put their demands for zero quotas on hold, and even the Japanese and Soviet governments, still perhaps in shock from the hugely different political environment they faced in the commission due to the influence now being exerted by environmental groups and public opinion, seemed more willing to cooperate in what promised to be a new era in cetacean management. Hopes of this nature, however, quickly evaporated as essentially the same uncertainty issues that previously had been raised in opposition to the Scientific Committee's majority opinions again were put forward in order to undermine confidence in the NMP, and therefore also in the Scientific Committee's capacity to sustainably manage whale stocks.

But the real challenge to the Scientific Committee's authority was less about knowledge gaps over biological data and stock numbers than it was about the fundamentally altered political priorities it was now working under (Heazle 2006). The main priority of a growing number of governments was no longer the exploitation of whale resources through sales of whale oil, but rather the exploitation of whales as endangered animals—and to some as sentient creatures. In societies with no economic or cultural whaling interests, doing so offered low-cost credibility as an environmentally responsible government while also avoiding the electoral disadvantage of going against domestic opinion opposing whaling (Heazle 2013; Kalland 1994; Friedheim 2001: 25).

Moreover, the Scientific Committee, which had unanimously opposed the US sponsored moratorium proposal, could no longer offer a consensus on either the NMP or the moratorium, since it was now divided between a majority, on one hand, who believed sustainable hunting, albeit at greatly reduced levels, to be possible under the NMP and, on the other hand, a small minority who regarded an ending of all commercial hunts as the only acceptable way of managing the uncertainties they raised in opposition to the NMP (Gambell 2001). According to Aron (2002), a member of the Scientific Committee during the NMP debate, opposition to the NMP, both in the commission and the Scientific Committee, did not emerge because of concerns over its scientific and practical integrity, but because of what it stood for and permitted, that is, a continuation of whaling endorsed by the IWC:

> The NMP, unlike the RMP, did not explicitly take errors into account in its formulation. However, in generating critical numbers, the low end of the estimates was selected to minimize the consequences of error. Also the terms of the NMP were, by themselves, very conservative. The big problem was less the data issue, which was solvable, but more the fact that the NMP permitted whaling. This is the real issue of the RMP, as well.

In the past, the Scientific Committee's divisions were limited to questioning either the need for lower quotas, or the extent to which quotas needed to be reduced, since the minority opinion traditionally had argued for higher catches against a majority view that most often favored precautionary reductions. Agreement within the Scientific Committee that some catches from some stocks were possible—except in cases such as with the blue and humpback stocks where the case for protection was very strong—had never before been in doubt, since none of the committee's scientists, regardless of their differences over catch limits, had considered a zero catch limit on all stocks of all species to be either justified or appropriate.

Despite the growing anti-whaling mood developing within the commission during the 1970s and 1980s, the issue of permanently banning commercial whaling still had not formally been raised in the IWC during this time. The moratorium itself was adopted in 1982 by the IWC as only a temporary pause until the concerns raised over scientific uncertainty were resolved—thereby implying, at this stage at least, that the commission felt it was possible to manage whaling safely. Accompanying the moratorium's entry to the Schedule to the Convention was a clause stating that:

> This provision will be kept under review, based upon the best scientific advice, and by 1990 at the latest the Commission will undertake a comprehensive assessment of the effects of this decision [i.e., the imposition of zero catch

> limits] on whale stocks and consider modification of this provision and the establishment of other catch limits.
>
> (IWC Thirty-Third Report 1983: 40)

Thus, the Scientific Committee had begun work on the comprehensive assessment programme on the assumption that the moratorium's main objective was the acquisition of more reliable cetacean research and management methods with the intention of considering a return to commercial whaling at some point. But as Greg Donovan (1989), the IWC Scientific Editor, noted, the original wording of the moratorium only stated that 'catch limits for the killing of whales for commercial purposes shall be zero' but before the vote was taken the extra wording including the comprehensive assessment and the 'establishment of other catch limits' was added. The extra wording was added, in Donovan's view, 'perhaps to indicate to the whaling countries that the proposal [the moratorium] seriously considered the possibility of whaling resuming' (Donovan 1989).

For several years after the moratorium's inception, and even during the negotiations prior to the vote being taken in the commission, the means by which the moratorium was to contribute towards more and better cetacean research remained unclear, with even the proposal for a comprehensive assessment of whale stocks being added as an apparent afterthought to the original moratorium proposal. Between the moratorium's adoption in 1982 and 1985, the Scientific Committee repeatedly told the commission that it did not understand what the comprehensive assessment was intended to mean and asked for an explanation and definition at each annual meeting. No answer from the commission, however, was forthcoming. In 1985 the Scientific Committee 'decided that if progress was to be made it [the Scientific Committee] would have to define what it thought was a "comprehensive assessment" and establish how it might be accomplished' (Donovan 1989). In April 1986, almost five years after the moratorium's adoption, a special meeting of the Scientific Committee was held in Cambridge, England, to develop a working definition of the Comprehensive Assessment and its aims.[12]

However, by 1986, when the Scientific Committee provided its definition of what the moratorium was intended to provide in terms of a comprehensive assessment, the IWC's management objectives already had become more confused than ever as the commission by this stage was openly divided over the interpretation of the ICRW. The two basic management objectives provided for by the ICRW are 1) that whale species are not hunted to extinction; and 2) that the maximum sustainable harvest is achieved. But as the IWC's Scientific Editor Greg Donovan has argued, these two objectives represent management extremes and require clarification in terms of how one is balanced against the other. According to Donovan (1995: 7), the 'setting of objectives and the relative weight given to those objectives (the trade-offs)' is a process that requires 'political rather than scientific decisions'.

Some governments that had supported the moratorium were now openly calling for a permanent end to commercial whaling on the basis of a recently developed interpretation of 'sustainable utilization' that excluded killing, while those who had opposed it continued to insist that hunting should resume after the comprehensive assessment's completion. Thus, the institutional environment in which the Revised Management Procedure—the major outcome of the Comprehensive Assessment—was to be developed could offer no consensus on why it was needed, or to what ends it would be used, if at all. Even the task of defining 'Comprehensive Assessment', which represented the reasoning behind the moratorium, was largely ignored by those supporting the moratorium who appeared content with simply proscribing commercial whaling as a solution to the commission's management problems, a strong indication of the political motivations underlying support for the moratorium.

## Science and the IWC 1993–present: gone, but not forgotten?

Within the IWC, an influential group of Western and South American states has remained opposed to *any* return to commercial whaling under IWC policy since the current moratorium's adoption in 1982. Their opposition is based on a position that highlights a clear tension between their commitment to science-based management on one hand, and the political influence exerted by anti-whaling non-government organizations (NGOs) and domestic electorates on the other (see, for example, Heazle 2013). When asked about the basis for the UK government's ongoing opposition to whaling at the 2001 IWC meeting in London, for example, the UK Commissioner Richard Cowan (2001) replied that 'Government policy on whaling is based on both science and public opinion. The fact is that the government is accountable to the electorate'.

The popularity and importance of whales within Western societies, as both a symbol for the fight against environmental destruction and also as a commodity to be enjoyed in unprecedented ways (i.e. whale watching, whale books, whale movies, whale songs), grew rapidly during the 1970s and 1980s, so much so that by the early 1990s, when the RMP and some elements of the Comprehensive Assessment were nearing completion, the status of 'the whale' was such that any research aimed at sustainable hunting was by now anathema in the eyes of the many governments and environmental NGOs invoking international and domestic public opinion to support what by now had become an entirely protectionist position on whaling. When the RMP finally was presented to the IWC in 1993 with the unanimous support of the Scientific Committee—a rare enough occurrence in itself—anti-whaling governments voted against its adoption, ostensibly on uncertainty grounds but mainly because of fears that its adoption would lead to an ending of the moratorium. Immediately following the commission's rejection of the RMP, the Scientific Committee's Chairman, Dr Philip Hammond, resigned as chairman in protest. In his letter of resignation to the IWC, Hammond (1993) wrote:

> What is the point of having a Scientific Committee if its unanimous recommendations on a matter of primary importance [the RMP] are treated with such contempt?. . . I have come to the conclusion that I can no longer

> justify to myself being the organizer of and the spokesman for a Committee whose work is held in such disregard by the body to which it is responsible.

By the time of the RMP's completion, many IWC scientists were, as Hammond's letter of resignation indicates, well aware of the decreasing relevance of their role in commission policy making. In response to the Scientific Committee's attempt to answer the concerns behind the moratorium by making management more reliable, whaling opponents responded by diluting science-based arguments over uncertainty issues with ethical arguments about inhumane killing methods and the immorality of killing allegedly highly sentient creatures such as cetaceans—arguments that identified their proponents with a protectionist rather than conservationist approach to wildlife management. By definition, a protectionist—as opposed to conservationist—position makes irrelevant any scientific evidence in support of sustainable hunting, which in effect means that even if uncertainty could be reduced to zero, which science cannot do, protectionists still would not accept any return to hunting. Driving this argument is the perception of whales as highly intelligent creatures that are different from other animals in addition to their status as a major symbol in the war against environmental degradation.

Thus, the idea of 'the whale', or what Arne Kalland (1994: 162–3) terms 'the Super Whale', became an effective rhetorical tool in down-grading the relevance of scientific research intended to create the type of sustainable whaling regime required by the ICRW (i.e. whales are special, therefore it is wrong to hunt them regardless of the level of scientific uncertainty involved). Uncertainty, meanwhile, continues to be invoked by protectionist-orientated opponents as a means of highlighting the kinds of risks a return to commercial whaling may produce—most recently in the context of climate change. The duality of this approach allows protectionist opponents—most notably Australia, New Zealand, United Kingdom and France, in addition to NGOs such as Greenpeace, the International Fund for Animal Welfare and the World Wide Fund for Nature—to keep one foot in the protectionist camp and the other in the conservationist camp in the course of unconditionally opposing whaling. The most likely explanation for this strategy is reluctance on the part of these governments and groups to divorce themselves entirely from the legitimacy offered by 'science-based' arguments, especially given the IWC's institutional requirement that policy 'be based on scientific findings'. According to Victor (2001: 298), uncertainty became 'a way to use the fig leaf of science to protect a moral argument'.

Douglas Butterworth (1992) and Arne Kalland (1994: 167) also have accused whaling opponents of duplicity in their use of scientific uncertainty. Kalland (1994: 167), for example, points out that science-based ecological arguments are 'more palatable for various reasons than are ethical and moral ones to a number of people, corporations, and government agencies' and argues that:

> many protectionists are more than reluctant to change their rhetoric from an ecological discourse to animal welfare or animal rights arguments. Thus

> the myth of the endangered whale is sustained by charging scientists producing new and larger stock estimates of being incompetent, biased and 'bought' by governments of whaling nations, by refusing to accept these new population estimates or refuting their relevance, or by introducing new arguments into the ecological discourse.

The RMP's eventual acceptance by the IWC in 1994 as an acceptable means of managing uncertainties in the available data quickly helped expose the extent to which conflicting views on the desirability of whaling per se, rather than only concerns for developing sustainable whaling, were driving disagreement and debate in the IWC (Heazle 2010; Victor 2001: 296–8). With the scientific dimension of anti-whaling policies (i.e. the need for more certainty) now undermined by the Scientific Committee's unanimous support for the RMP, whaling's opponents were forced to rely less on the 'fig leaf of science' and more on the political legitimacy that could be drawn from the moral and ethical foundations of their opposition. Indeed, by the time of the RMP's completion and adoption by the IWC in 1994, some members had even gone so far as to say the management uncertainties that had become almost the exclusive focus of IWC debate over the previous 20 years were now not even relevant to the question of whether or not commercial whaling can recommence. In its opening statement to the IWC's 1994 meeting, the New Zealand government stated its intention to participate in negotiating the rules under which the RMP could be implemented (i.e. the Revised Management Scheme—RMS) even though its opposition to whaling would continue:

> We will work to maintain the moratorium on commercial whaling because it reflects the current reality of world opinion. We will participate fully in the dialogue about the Revised Management Scheme because it is essential to have the best possible rules for whaling, whether or not they are required in practice.
>
> (New Zealand Opening Statement, IWC 1994)

At the same meeting, the Australian Commissioner stated that: '[W]e remain strongly opposed to all commercial whaling . . . such activities are no longer needed to meet essential human needs. . . .'

In extreme cases such as in the late 1950s and early 1960s, when rapidly declining pelagic catches convinced whaling governments that many stocks had been severely depleted, the Scientific Committee's advice to reduce catch limits was eventually accepted because of broad political acceptance of a recognized certainty—rather than uncertainties. But even in such cases, political agreement on the issue (i.e. species should not be driven to extinction) was still required before scientific advice was able to directly influence the commission's policies. No political consensus on the desirability of the RMP as a means of actually managing commercial hunting existed at any time during its development and, despite its belated adoption by the commission in 1994, anti-whaling governments continue to oppose its implementation.

## Conclusion

The Scientific Committee's very limited influence on IWC policy over most of the last six decades demonstrates how long the role of EA as a source of policy legitimacy can remain on the fringes of policy making in institutional environments where technical uncertainties, the range of both national and transnational policy actors involved, and the variety of values and policy interests they pursue, expand well beyond what the institution's original arrangements were intended to manage. Due to the 'crisis of legitimacy' created out of the IWC's institutional inability to better manage the scientific and political complexities posed by an increasingly varied and often conflicting array of ideas about what 'whaling' represents, the relevance of expert findings to IWC policy, and which findings should be regarded as 'authoritative', the influence of the Scientific Community—the commission's appointed source of expertise—over the last 60 years has been extremely limited. The IWC experience as a 'science-based' management regime suggests that during times of strong opposition to the expert advice on offer (due to the interests its acceptance is likely to compromise—the 1950s), or increasing division over the ends it is intended to serve (managing hunting versus preventing hunting—1970s to the present), science, and the authority it invokes, is unlikely to be able to guide political agreement on policy means while the much more fundamental dispute over the ultimate ends IWC policy is intended to serve continues. When disagreement over policy ends becomes protracted and beyond negotiation, as it has in the IWC from the mid-1980s, science and the expert authority it invokes effectively has no role to play in policy debate—regardless of any effort to strengthen its authority through better science—other than as handmaiden to the political viewpoints its findings appear to support.

EA's formal role as a key source of policy legitimacy in the IWC most often has been undermined by uncertainty arguments invoked by different actors at different times, depending on the fit between their national interest perceptions and the expert advice on offer from the Scientific Committee. When uncertainty arguments became more difficult to sustain following the commission's adoption of the RMP in 1994—a system specifically designed to manage irreducible uncertainties in catch quotas—governments opposed to whaling resorted instead to moral arguments against any killing of whales regardless of whale numbers and management provisions. The implication here for EA in the IWC is that under this view scientific findings and advice that makes possible or involves the killing of whales, such as the RMP, becomes irrelevant to policy despite its ongoing institutional relevance under the ICRW; science promoting and making possible the non-lethal utilization and investigation of whale stocks thus becomes the only legitimate form of EA for anti-whaling advocates. What this means, in effect, is that the credibility and relevance of EA in the IWC has reached a point where it now is almost entirely contested on political rather than scientific grounds, thereby leading to the blurring of the two forms of authority and subsequent deadlock over what legitimate policy should represent.

As numerous studies (Jasanoff 1990; Sarewitz 2004; Hoppe 2005; Fischer 2009) demonstrating the difficulties of isolating expertise and evidence from political influence, especially when expertise enters the policy process (Collingridge and Reeve 1986), have shown, the Scientific Committee's contribution to policy making was always going to be tempered to some degree by more than only 'scientific findings'. But, as the major policy debates in the IWC examined here indicate, the policy justifications adopted by various governments often have reflected more than only the now broadly recognized problems that prevent expertise from 'objectively' informing policy. Indeed, almost all the commission's most important management policy decisions to date have been based on arguments either attacking or ignoring the Scientific Committee's conclusions and advice, leaving many of the commission's scientists resigned to only proposing what they thought governments would accept—as occurred in the 1950s; or sidelined because the issues they felt able to give advice on (e.g. managing uncertainty in limited hunting) were no longer relevant to the 'new' rules and principles for IWC policy being pushed by anti-whaling governments and groups.

The Scientific Committee, lacking the authority to unilaterally reinterpret its role under the convention, has necessarily remained tied to the duties and responsibilities given to it by the convention. However, governments, able to exercise the rights they enjoy under international law as the representatives of sovereign states, have instead reinterpreted, and at times directly challenged, the institutional rules set down by the ICRW as the basis of their policy positions.[13] The result, as I have argued here, has been the creation of an 'institutional void' in which the convention's goals, principles, and rules are not commonly understood for the purposes of policy; the ensuing absence of any common political direction and purpose among the IWC's members has, as a further consequence, made the role of science and expertise in commission policy making, as set out by the ICRW, at best contested, and at worst irrelevant.

Deciding an appropriate balance between science-based understandings of, and requirements for, sustainability (i.e. catch quotas that do not deplete stock numbers), what sustainability *should* mean (e.g. some hunting versus no hunting), and which interests should be prioritized under varying levels of uncertainty all require essentially political rather than only scientific judgement. So it is unsurprising that as political attitudes towards whaling changed from the 1970s onwards, this initial tension, which in practice had been earlier resolved in favor of industry interests at the expense of sustainable catches until the inevitable collapse of the Antarctic stocks in the early 1960s, then re-emerged as a more fundamental conflict over what the ICRW's *raison d'être* should be—that is, whether any commercial, and later even lethal scientific sampling, regardless of the available scientific advice on offer, is acceptable.

In the absence of an institutional environment where broadly supported policy objectives can at least be identified and maintained, scientists hardly can be expected to provide relevant policy advice. The inability (or unwillingness) of the IWC membership to provide this direction by collectively reforming the IWC's objectives and operating rules are key to understanding the very limited influence of the Scientific Committee's work in the past and its irrelevance today. The only exception to the political division that has come to characterize the IWC was a short period during from the mid 1960s until the early 1970s when political consensus on the commission's ends and how they should be achieved briefly allowed the Scientific Committee to play the role envisaged for it by the ICRW. Indeed, expert authority in the IWC always has been important insofar that the commission's convention requires that policy at least appear to be based on 'scientific findings'. As an input to policy, however, the findings and advice of the Scientific Committee—despite the significant role given to the authority of experts by the ICRW—have in many cases been at best disputed and at worst ignored within the IWC's ongoing institutional void.

## Notes

1 The preamble to the 1946 Convention, ratified by 12 governments by the time of the IWC's first meeting in 1949, states 'the history of whaling has seen overfishing of one area after another and of one species after another to such a degree that it is essential to protect all species of whales from further over-fishing'. This goal is to be achieved, the Convention further states, by allowing whale stocks to return to their 'optimum' levels 'as rapidly as possible without causing widespread economic and nutritional distress' (IWC First Report 1950: 9).

2 See, for example, Steinar Andresen's 2001 assessment of the IWC's institutional arrangements, which he argues are appropriate; the problem, Andresen argues, is that improvements in the 'scientific process' have been 'upset to a large extent by the strong infusion of politics and conflicts in the Scientific Committee' (see Andresen 2001: 240).

3 See, for example, Arne Kalland (1994) and Charlotte Epstein (2006) for contrasting explanation and analysis of the significance of 'the whale' as a 'symbol for collective action' in response to the emerging political salience of environmental issues during the late 1960s and early 1970s.

4 The IWC has not managed whaling since the moratorium went into effect in 1986. Whaling has legally continued outside the limits imposed by the moratorium among some members including Japan, Norway, and Iceland, but has not been subject to IWC control.

5 'to provide for the proper conservation of whale stocks and thus make possible the orderly development of the whaling industry' (IWC First Report 1950: 10).

6 Japanese whaling remained profitable due to strong domestic demand for both whale meat and whale oil. The USSR industry did not need whaling to be profitable because of the generous level of investment provided by the Soviet state.

7 See IWC Seventh Report (1956: 18) for delaying and eventual withdrawal of Scientific Committee request for conference funding in 1954; and also IWC Second Report (1951: 6) for the commission's rejection of funding request for publication of Scientific Committee reports.

8 The blue whale had the highest yield and therefore became the standard against which other whales were measured (i.e. 1 blue whale unit = 1 blue whale = 2 fin whales = 2.5 humpback = 6 sei whales).

9 Dutch scientists E.J. Slijper and E.F. Drion became well-known opponents within the Scientific Committee to quota reductions during the 1950s. Both scientists, conscious of the economic pressure the state-funded Netherlands fleet was under, regularly opposed proposals for quota reductions in the Scientific Committee on the grounds

of uncertainty and alternative interpretations of the available evidence. These arguments made it possible for opposition to reductions to at least appear to be grounded in science rather than self-interest. See Schweder (1992) and Heazle (2006).

10 At the 1971 meeting, the Scientific Committee informed the commission that it had been 'unable to reach a single estimate for the sustainable yield [of the Antarctic fin whale stock] in 1971/72. Most members of the Committee believed that the best estimate was 2,200 whales (1,100 blue-whale units), while the Japanese scientists estimated a yield of 4,250 (2,125 blue-whale units) with which the USSR scientists agreed'.

11 In contrast to Australia's later support for the moratorium and its current opposition to any return to commercial whaling, the Australian government argued when proposing the NMP on: 'the need to preserve and enhance whale stocks as a resource for future use and taking into consideration the interests of consumers of whale products and the whaling industry as required by the International Convention on Whaling' (IWC Twenty-Sixth Report 1976: 25).

12 The IWC report of that meeting defined the Comprehensive Assessment as: 'an in-depth evaluation of the status of the stocks in the light of management objectives and procedures. This could include examination of current stock size, recent population trends, carrying capacity and productivity.' See 'Report of the Special Meeting of the Scientific Committee on Planning for a Comprehensive Assessment of Whale Stocks' in Donovan (1989: 3).

13 See, for example, Australia's recent case in the International Court of Justice against Japan's right under the ICRW to conduct lethal scientific whaling research.

## References


Allen, K.R. 1980. *Conservation and Management of Whales*. Seattle, WA: University of Washington Press.

Andresen, S. 2001. 'The Whaling Regime: "Good" Institutions but "Bad" Politics?' In *Toward a Sustainable Whaling Regime*. R.L. Friedheim (ed.). Seattle, WA: University of Washington Press, 235–65.

Aron, W. 2001a. 'Science and the IWC.' In *Toward a Sustainable Whaling Regime*. R.L. Friedheim (ed.). Seattle, WA: University of Washington Press, 105–22.

Aron, W. 2001b. 'Interview with Author.' Tokyo, March.

Aron, W. 2002. 'Email correspondence with Author.' January.

Australia Opening Statement. 1994. Given to the Forty-Sixth Annual Meeting of the International Whaling Commission. Puerto Vallarta, May.

Bailey, J.L. 2008. 'Arrested Development: The Fight to End Commercial Whaling as a Case of Failed Norm Change.' *European Journal of International Relations*, 14(2): 289–318.

Bannister, J. 2001. 'Interview with Author.' London, July.

Beverton, R.J.H. and Holt, S.J. 1957. *On the Dynamics of Exploited Fish Populations*. Dordrecht: Springer Science and Business Media.

Birnie, P. 1985. *International Regulation of Whaling: From Conservation of Whaling to Conservation of Whales and Regulation of Whale Watching*, vols 1 and 2. New York: Oceana Publications.

Butterworth, D. 1992. 'Science and Sentimentality.' *Nature*, 357: 532–4.

Collingridge, D. and Reeve, C. 1986. *Science Speaks to Power: The Role of Experts in Policy Making*. London: Pinter.

Cowan, R. 2001. 'Interview with Author.' London, July.

Cushing, D.H. 1988. *The Provident Sea*. Cambridge: Cambridge University Press.

Donovan, G.P. (ed.) 1989. *The Comprehensive Assessment of Whale Stocks: The Early Years*. Special Issue 11. Cambridge: International Whaling Commission.
Donovan, G.P. 1995. 'The International Whaling Commission and the Revised Management Procedure.' In *Additional Essays on Whales and Man*. E. Hallenstvedt and G. Blichfeldt (eds). Lofoten, Norway: High North Alliance.
Epstein, C. 2006. 'The Making of Global Environmental Norms: Endangered Species Protection.' *Global Environmental Politics*, 6(2): 32–54.
Fischer, F. 2009. *Democracy and Expertise: Reorienting Policy Inquiry*. Oxford: Oxford University Press.
Friedheim, R.L. 2001. 'Introduction: The IWC as a Contested Regime.' In *Toward a Sustainable Whaling Regime*. R.L. Friedheim (ed.). Seattle, WA: University of Washington Press, 3–48.
Gambell, R. 1999. 'The IWC and the Contemporary Whaling Debate.' In *Conservation and Management of Marine Mammals*. J.R. Twiss Jn. and R.R. Reeves (eds). Washington, DC: Smithsonian Institution Press, 179–97.
Gambell, R. 2001. 'Interview with Author.' London, July.
Hajer, M. 2003. 'Policy Without Polity: Policy Analysis and the Institutional Void.' *Policy Sciences*, 36(2): 175–95.
Hajer, M. 2009. *Authorative Governance: Policy Making in the Age of Mediatization*. Oxford: Oxford University Press.
Hammond, P. 1993. 'Letter of Resignation to the IWC.' IWC/2.1, May 26.
Heazle, M. 2006. *Scientific Uncertainty and the Politics of Whaling*. Seattle, WA: University of Washington Press.
Heazle, M. 2010. *Uncertainty in Policy Making: Values and Evidence in Complex Decisions*. London: Earthscan.
Heazle, M. 2013. 'See You in Court!: Whaling as a Two Level Game in Australian Politics and Foreign Policy.' *Marine Policy*, 38: 330–6.
Hoppe, R. 2005. 'Rethinking the Science-Policy Nexus: From Knowledge Utilization and Science Technology Studies to Types of Boundary Arrangements.' *Poiesis Prax*, 3(3): 199–215.
IWC. 1950. 'First Report of the Commission [1949].' London.
IWC. 1955. 'Sixth Report of the Commission [1954].' London.
IWC. 1956. 'Seventh Report of the Commission [1955].' London.
IWC. 1960. 'Eleventh Report of the Commission [1959].' London.
IWC. 1961. 'Twelfth Report of the Commission [1960].' London.
IWC. 1964. 'Fourteenth Report of the Commission [1962].' London.
IWC. 1970. 'Twentieth Report of the Commission [1968].' London.
IWC. 1971. 'Twenty-First Report of the Commission [1969].' London.
IWC. 1973. 'Twenty-Third Report of the Commission [1971].' London.
IWC. 1974. 'Twenty-Fourth Report of the Commission [1972].' London.
IWC. 1975. 'Twenty-Fifth Report of the Commission [1973].' London.
IWC. 1976. 'Twenty-Sixth Report of the Commission [1974].' London.
IWC. 1983. 'Thirty-Third Report of the International Whaling Commission [1982].' Cambridge.
Jasanoff, S. 1990. *The Fifth Branch: Science Advisers as Policymakers*. Cambridge, MA: Harvard University Press.
Kalland, A. 1994. 'Whose Whale Is That? Diverting the Commodity Path.' In *Elephants and Whales: Resources for Whom?* M.R. Freeman and U.P. Kreuter (eds). Amsterdam: Gordon and Breach, 159–86.

New Zealand Opening Statement. 1994. Given to the Forty-Sixth Annual Meeting of the International Whaling Commission. Puerto Vallarta, May.

Sarewitz, D. 2004. 'How Science Makes Environmental Controversies Worse.' *Environmental Science & Policy*, 7(5): 385–403.

Schweder, T. 1992. 'Intransigence, Incompetence or Political Expediency? Dutch Scientists in the International Whaling Commission in the 1950s: Injection of Uncertainty.' IWC Scientific Committee Paper 44/013.

Tønnessen, J.N. and Johnsen, A.O. 1982. *The History of Modern Whaling*. London and Canberra: C. Hurst & Company and Australian National University Press.

Victor, D. 2001. 'Whale Sausage: Why the Whaling Regime Does not Need to Be Fixed.' In *Toward a Sustainable Whaling Regime*. R.L. Friedheim (ed.). Seattle, WA: University of Washington Press, 292–310.

# 5 On the interdependency of political authority and economic expertise

*John Kane*

It is a central contention of this volume that political authority in a liberal democracy comes in the first instance from electoral success, but secondarily and very importantly from continuing success in policy and performance. One of the most fundamental tasks of government is 'managing the economy', economic failure being practically synonymous with political failure.[1] This makes political leaders, who form the essential hinge linking democratic polity with liberal economy, especially dependent on a particular class of advisors whose authority rests on the perceived validity and application of the economic theories they espouse.

It is the argument of this chapter that this fact creates a very special interdependency between the authority of economic experts and the authority of political leaders. When economies prosper, both the prestige of economic experts and the authority of leaders are strengthened; when economies struggle, the perceived authority of each tends to fail. It is also observed, however, that the social attribution of expert authority is not uniformly applied across economists, whose views and prescriptions vary markedly, but generally to a subset that has captured the policy field at some particular time. This is a political as much as an intellectual achievement, and a necessary one if economists hope to be heard. Expert authority, in other words, is a function of consensus formation among political leaders as to whose policies offer the best hope of helping them in their economic management task or a way out of current difficulties.

In the wake of the Global Financial Crisis (GFC) and the recession and stagnation that ensued, the intellectual authority of a whole class of neoclassical economists was severely dented, with consequences also for the authority of political leaders subject to their advice. Insofar as those economists had claimed the status of bona fide scientists, the validity of their 'science' came into serious question, particularly with regard to the political commitments underpinning their complex mathematical modeling. However, since all policy prescriptions regarding economic management must rest on *some* view of the essential nature and operation of a modern capitalist (now global) economy, the question was to whom political leaders could now turn to find a way back from chronic stagnation to general prosperity. As noted, alternate versions of economics had long been around, but none had acquired the kind of authority that comes from the formation about them of a political consensus, and thus none were politically available for leaders who tended to cling to now discredited shibboleths for want of anything better. With serious imbalances in the global economy effectively unaddressed, the danger was that the world's political leaders would collectively stumble toward an even greater crisis without either the intellectual or practical means to counter it.

## Democratic polity, liberal economy and economic crisis

I have argued elsewhere (Kane 2014) that democratic leaders face particular problems in performing the essential task of economic management. Among other things, they must understand the complexities of a modern capitalist economy with its decentralized markets in which a myriad of agents with hugely different dimensions of power and influence—households, businesses, financial institutions—make repeated decisions about employment, income, consumption, investment and borrowing often under conditions of considerable uncertainty. A leader with management responsibility for this complex system would seem to have to be an economist, yet democratic leaders are seldom economists and therefore are inevitably peculiarly reliant on expert advisors.

The actions of political leaders will not necessarily always prove congruent with the advice of economic experts, because the former are forced to respond to many conflicting pressures. Indeed, it is a common complaint of economists that vulgar 'politics' too often trumps their rational policy recommendations. If, for example, Friedmanite monetary policy failed to control inflation in Western nations in the 1980s, this was not a consequence of a flaw in the theory but of the failure of governments to properly control the money supply due to political factors. But this perennial struggle between 'rational' economic policy and political imperatives merely reflects the structurally critical but nevertheless anomalous position that the democratic leader occupies as the essential hinge linking a democratic polity with a liberal economy. In their democratic role, leaders are agents and representatives of the people, concerned first and foremost with safeguarding the people's welfare and democratically responding to their demands; but as economic managers leaders must be solicitous of an economic system whose fundamental principle is founded not in democratic theory but in market liberalism.[2]

Leaders generally believe, of course, that fostering a healthy economy (growing the pie) is the surest way of fulfilling their democratic responsibilities. This indeed was the post-Depression, post-Second World War solution to the problem of balancing economy and democracy, one that was said to take the form of an implied 'social contract': governments, through judicious control over the essential parameters of international trade and through 'Keynesian steering' of domestic economies, could simultaneously satisfy the demands of business for profitability and those of the mass of people for income through regular employment, and increasingly more by virtue of growth.[3] This, of course, meant continued reliance on a dynamic system of capital accumulation that was extremely productive but also highly volatile and prone to cycles of boom and bust, in the latter stage of which profitability collapsed and unemployment rose sharply, though this problem seemed solved by Keynesian fiscal policies in the post-war period. In the 1980s, however, Keynesian demand management failed in circumstances of 'stagflation'. In response to this crisis, the post-war social contract was unilaterally rewritten by political leaders of whatever ideological stripe and the balance tipped from democratic control toward economic liberalism, in the hope that resultant prosperity would anyway satisfy democratic demands.

So long as prosperity seemed to revive as a consequence, the democratic requirement could be argued to be fulfilled. But in times of economic trouble, and especially in the wake of the GFC, the tensions within the leadership role became acutely apparent. The measures they felt obliged to take on the recommendation of expert advisors to heal an ailing economy often seemed to conflict with the demands or interests of a democratic polity. Thus, in the immediacy of the economic crisis of 2008, first President Bush and then President Obama were caught in the classic dilemma of trying to serve the people (Main Street) by helping those who seemed to have harmed them (Wall Street). Seven years later, Western economies still languished while governments suppressed interest rates and printed money, sustaining banks and inflating stock markets with minimal discernible effect on the 'real' economy. Even in the US where a fragile recovery seemed underway, workers' wages remained flat. In Europe, where financial crisis had turned into a political crisis of the European Union, a series of ad hoc measures averted outright disaster while leaving vast structural problems unaddressed, perpetuating inequalities and deferring decisive action. Meanwhile, the German-backed imposition of 'austerity' on Europe as a remedy for economic malaise seemed, especially to the long-suffering peripheral countries, self-defeating, causing great resentment whose eventual fruit was the electoral victory of the radical leftist Syriza party in Greece.

Policy debate over fiscal stimulus versus austerity was a marked feature of the crisis, a debate that was vitiated by what seemed the equally unpalatable consequences of each—either mounting deficits or damaging deflation. Both positions appeared, moreover, to rest on rather impoverished intellectual grounds. Stimulus was indelibly associated with Keynesian demand management, generally believed to have been discredited during the 1970s–1980s era of 'stagflation', while austerity harked back to an era before the Keynesian moment, the 1930s, when it was imposed during an economic crisis that subsequently turned into deep and lasting depression with well-known political consequences (Blyth 2013). The post-war Keynesian consensus of Western countries that lasted till the 1970s had famously (or notoriously) been displaced by an even older strand of thought, revived free market, laissez-faire liberalism, now rebranded neoliberalism, but this itself appeared to come to grief after being applied too liberally to financial markets in the era of 'the Great Moderation'.[4]

Financial crisis and ensuing economic stagnation caused, or perhaps exposed, disarray among economists that left political leaders at the mercy of forces they could only very inadequately grasp. Certainly, there was widespread critique of the neoclassical economists whose work had underpinned the neoliberal policies of liberalization and deregulation, and who were variously blamed for not having predicted the crisis, for having helped cause it, and for being unable to address it. The authority of these economists, and of the political advisors whose claim to expertise generally relied on their operating premises, had rested practically on the perceived success of neoliberal policies and intellectually on the authority provided by the prestige of the neoclassical discipline, which they typically claimed to be a 'science' and even the only genuine science among the social sciences. After 2008 there was renewed scrutiny of this claim and of the barely concealed political premises informing it.

## Science and ideology: the failure of neoclassical modeling

Ideas are always central to policy but whether and how particular ideas get taken up by political actors and become (or fail to become) embedded in institutional structures and behaviors is always a complicated matter of political action, advocacy and competition in combination with elements of chance (Kingdon 1995). This is no doubt true for all areas of human endeavor, but it presents special difficulties in the field of economics. It is usually assumed, on any reasonable scientific model, that ideas that survive vigorous contestation do so as a result of both internal consistency and the explanatory power they display when tested against the world. Bias, prejudice, vested interest there will always be in human affairs—elements we may describe as 'political' in Weber's (2004 [1919]) sense of the political as defined by *ira et studium* (anger and bias)—but these passionate biases are expected to be ironed out in the long run by the arbiter of empirical evidence. Theories may succeed, perhaps only always provisionally, but they can fail decisively. Economic thought, however, is so intertwined with, and implicated in, the political that the long run can be a very long time in coming.

The irony is that neoclassical economics conceived itself as, in significant fashion, a discipline to be clearly distinguished, if not quite divorced, from the political. The desire to raise economics to the level of a 'science' in the manner of the physical sciences required that there exist a distinctive subject matter susceptible to rigorous mathematical investigation and analysis.[5] That independent matter was the 'naturally' free market. Like the liberal theory from which it descends, free market economics presumes asocial, pre-political individuals for whom artificially constructed government is as much a threat to natural freedom as a solution for its inconveniences. However useful or necessary governments are for establishing and maintaining the rule of law,[6] laws of property and contracts and so forth, their interventions remain those of an external power capable of facilitating the operations of or removing obstacles to free markets but not intrinsically foundational to their creation. More problematically, governmental interventions may form one of the chief impediments to thriving markets. A necessary element of power relations is thus reluctantly admitted, sitting uncomfortably within a vision of individualized human agents freely producing and freely exchanging in a market that efficiently allocates resources and fairly apportions rewards.

This is a highly idealized, not to say utopian vision. Liberal economists, building on Smith's invisible hand,[7] assumed that, beneath the variety of social-historical forms, there lurks a given set of acquisitive human desires, which, let freely reign, naturally express themselves in exchange relations geared to harmoniously mutual benefit.[8] Liberalism fosters a basically anti-political outlook by depicting the state as a perennial threat to this 'natural' economics. Mark Blyth terms the resultant ambivalent attitude of liberal economics toward the state as 'can't live with it, can't live without it, don't want to pay for it' (Blyth 2013: 98). The origins of this attitude lie in an era when a nascent commercial civilization was struggling to free itself from the inhibitions of medieval custom, the irksome demands of capricious monarchs and the mercantilist restrictions of governments to establish new conditions of production and exchange. In our developed and complex capitalist world such an era is a distant memory, but its enduring influence can be seen in the evergreen political issue of the proper relationship between state and economy. Dominant political ideologies are located at some point along the continuum from hostility to the state to necessary interdependence to state dominance with neoclassical economics generally positioning itself at the minimalist state end.

The pretensions of neoclassical economics to become a hard, objective science—implying *sine ira et studium*—thus did not preclude political positioning, nor prescribing policy for governments. Indeed, its political agenda was set by that science's founding presuppositions. Benign but vulnerable markets must be simultaneously defended by and from government. Free market economists wished to use political authority to preserve and extend the domain of market freedom, but they also wished to persuade political authority to exercise restraint in intervening in markets. A typical example was provided by Milton Friedman when he expressed a longing for 'automatic systems' that would put his beloved monetary policy beyond the reach of political manipulation (cited in Turner 2007: 77).[9]

More problematic for neoclassical 'science' was that its over-simple view of natural, virtuous market exchange encouraged the use of abstract, oversimplified and often highly unrealistic assumptions in their analytical modeling—most foundationally, 'the combined assumptions of maximizing behavior, market equilibrium, and stable preferences, used relentlessly and consistently', as Gary Becker (1976: 5) observed. True, this simplification rendered their subject matter susceptible to increasingly elaborate and complex mathematical econometric modelling, bringing undoubted prestige to economics and allowing economist Edward Lazear (2000) to write triumphally that economics was not just a social science but a 'genuine science', allowing it imperialistically to invade intellectual territory previously believed beyond its realm (for example, political science).

Nevertheless, neoclassical models have often been criticized for their unrealistic assumptions, as well as for their lack of predictive power and their uselessness for informing policy. The modelers have by and large fallen back on a justification provided in a famous essay from 1953 by Friedman, 'The Methodology of Positive Economics', which seemed to license the use of 'assumptions' that bore ever smaller resemblance to reality. There is, however, a considerable irony in this use of Friedman whose basic intentions were more conventionally 'scientific'.

This can be seen from recent scholarship on the epistemological status of economic models and of scientific models generally, which describes them as either *isolationist* or *fictionalist* (Grüne-Yanoff 2009). Isolationists regard modelling as a way to isolate causal factors or capacities of the real world; fictionalists regard models as parallel fictional worlds, populated by fictional agents and fictional institutions, and possibly operating under fictional forces and principles, from which one may be justified under certain conditions in making inferences to the real world.[10] But the fact is that most non-economic science models are isolationist. A scientific theory proposes a principle about the way the world, or some part of it, actually works, that requires confirmation or disconfirmation through empirical testing. Mathematization, especially for the physical sciences, is a tool—a powerful and necessary one—used to formalize explanatory theories in a way that allows the deduction of various consequences for testing. It is one part of a complex methodology combining induction, deduction and experimentation (and usually a good deal of creative intuition).[11]

A sympathetic re-reading of Friedman's influential essay shows that his prime intention was directed at just such an isolationist model, but certain false conclusions he drew in the process of argumentation ironically gave heavy impetus to the fictional one. He wrote: 'The ultimate goal of a positive science is the development of a "theory" or, "hypothesis" that yields valid and meaningful (i.e. not truistic) predictions about phenomena not yet observed' and 'the only relevant test of the validity of a hypothesis is comparison of its predictions with experience' (Friedman 1953: 7, 12). But because appearances are deceptive, evidence must be interpreted and organized to reveal simple structures underlying complex reality. But such theoretical penetration, said Friedman, requires the use of simplifying 'assumptions', and his discussion of these left an unfortunate legacy. He asserted:

> Truly important and significant hypotheses will be found to have 'assumptions' that are wildly inaccurate descriptive representations of reality, and, in general, the more significant the theory, the more unrealistic the assumptions (in this sense). The reason is simple. A hypothesis is important if it 'explains' much by little, that is, if it abstracts the common and crucial elements from the mass of complex and detailed circumstances surrounding the phenomena to be explained and permits valid predictions on the basis of them alone. To be important, therefore, a hypothesis must be descriptively false in its assumptions.
>
> (Friedman 1953: 14)

But abstracting 'common and crucial elements' does not assume these are 'unrealistic' or 'descriptively false'—quite the contrary. In a later passage Friedman claimed that a completely 'realistic' theory of, say, a wheat market would have to include not just factors of supply and demand but the kinds of credit instruments used, the personal characteristics of the traders, color of their hair and eyes, and so on ad infinitum (Friedman 1953: 32). Clearly, what he intended was to abstract real and relevant factors from others irrelevant to economic explanation. His actual account licensed other theorists' use of 'assumptions' that bore ever smaller resemblance to reality. The result was that 'science' in neoclassical economics became wholly and unqualifiedly identified with 'modeling' per se, and thus hardly scientific at all.[12]

According to recent critics, neoclassical theories have survived, despite their inherent weaknesses, not as objective science but as ideology specifically allied to neoliberal ideology (Slattery *et al.* 2013). The prestige conferred by the elaborate appearance of hard science served to occlude the fact that at their heart lay the optimistic political (anti-politically political) assumption of the natural, equilibrially self-adjusting (i.e. intrinsically good) market. This in practice could be dangerously misleading. The main manifestation of it leading up to the GFC was overreliance on the so-called Efficient Market Hypothesis (EMH), which states that the price of a financial asset reflects all available information relevant to its value.[13] The implication is that investors cannot consistently achieve returns in excess of the average, because an undervalued asset would be immediately bought by informed investors, and an overvalued one quickly sold short thus preventing asset bubbles forming, or from persisting, as observant investors would spot them and make money by popping them.[14] Alan Greenspan noted that this paradigm 'was so thoroughly embraced by academia, central banks, and regulators that by 2006 it had become the core of the global bank regulatory standards known as Basel II' (Greenspan 2013). The EMH informed and inspired financial engineering on Wall Street that produced increasingly complex derivatives and securitizations, credit default swaps and collateral debt obligations that supposedly reduced risk by spreading it while in reality dangerously heightening it.

Nobel prizewinner Robert Lucas, whose work had strengthened the intellectual acceptability of the EMH, contemptuously rejected post-crisis critique of the models based on it, arguing in effect that econometric models could not be expected to predict what they were not designed to predict (crises). That this might represent a fatal weakness in a hypothesis that presumes the impossibility of prolonged asset bubbles—of the kinds that throughout history have preceded financial collapse (Reinhart and Rogoff 2009)—was not considered by Lucas (2009). Implicit but unstated in his defense was the assumption that the fall of Lehman Brothers was an unforeseeable exogenous shock, no doubt the consequence of a poor public decision, therefore not the responsibility of economists and their models. Such a view was quite in conformity with an economic theory whose central doctrinal element of self-equilibrating markets makes all shocks to the system exogenous by definition.

Other simplifications, if not outright dangerous, were at least impediments to understanding the real dynamics of a complex modern economy. For example, the dedicated neoclassical focus on individual exchange (essentially of a bartering kind) fostered a general neglect of institutions, not just political ones but commercial and financial ones as well (even the firm is treated as a quasi-individual with a single will guided by rational economic choice). Lucas again provides a telling example of the shortcomings of such a vision. In an article on the 'mechanics' of economic development, he stated:

> I will also be abstracting from all monetary matters, treating all exchange as though it involved goods-for-goods. In general, I believe that the importance of financial matters is very badly over-stressed in popular and even much professional discussion and so am not inclined to be apologetic for going to the other extreme. Yet insofar as the development of financial institutions is a limiting factor in development more generally conceived I will be falsifying the picture, and I have no clear idea as to how badly.
>
> (Lucas 1988: 6)

Of course, this falsification turned out to be fatal when it came to the 'mechanics' of the GFC, which could scarcely be understood if intermediating financial institutions were excluded.[15]

A robust political economy adequate to the character of modern capitalism could hardly be built on such foundations, but then neoclassicists did not pretend to be doing political economy, just economics. The question was why the apparent wreck of their scientific authority in the financial crisis did not produce one of those vaunted 'paradigm shifts' that failure is supposed to provoke. Myron Scholes—another Nobel prizewinner and co-creator of the famous Black–Scholes mathematical model for pricing stock options—argued that, despite the apparent failures revealed by the financial crisis, the efficient market paradigm was not dead, because: 'To say something has failed you have to have something to replace it, and so far we don't have a new paradigm to replace efficient markets' (cited in *The Economist* 2009).

Nevertheless, the question remained as to why authority on economic matters did not move to a more plausible source. There had always been schools of economists labeled 'heterodox' (tellingly, perhaps)[16]—Austrians, neo-Keynesians, post-Keynesians, behavioralists, Marxists—critiquing the neoclassicals from the academic fringes and many of these certainly received renewed attention in the aftermath of crisis. What they had in common, for all their differences, was a refusal to accept the neoclassical sidelining of attention to matters of distribution and power. Their identifying core was 'the view that capitalism cannot be even approximately described by an autarchic equilibrium model of market processes' (Dymski 2014: 2). In other words, a valid economics must also be a political economics. But however plausible and more realistic might be the variant analyses of heterodox economists, none had made much headway during the neoliberal era. Nor was any sufficiently poised to capture the field and establish its authority in a politically effective manner when the GFC brought neoliberal orthodoxy into serious question. What creating this political groundwork might have taken can, in fact, be illustrated by the example of the neoliberal movement from the 1940s onward, as can also the void left by the latter's crisis.

## Formation and downfall of the neoliberal consensus

The turn toward Keynesian macroeconomic thinking in the 1940s had occurred in the context of long-running economic malaise and international conflict. The turn toward neoliberal theory also occurred in the context of crisis, but the political ground for its entry had been long under preparation. The general 'collectivism' of the inter-war and post-war era had been intellectually resisted by liberal thinkers from the 1930s, and most significantly by members of the Mont Pèlerin Society whose founders in 1947 included Friedrich Hayek, Milton Friedman, Ludwig von Mises, Karl Popper and Michael Polanyi (Hartwell 1995; Mirowski and Plehwe 2009). Despite their general adherence to free market economics, the Mont Pèlerin community made no secret of the fact that theirs was a resolutely political movement. In quasi-Marxist fashion, they assumed that the economic base determined the political superstructure. Thus, economic freedom based on private property and the competitive market was an essential base for a politically free society, the preserver of the individual freedoms central to the human dignity they saw threatened by the arbitrary power of intolerant creeds.

European neoliberalism prior to Mont Pèlerin was not necessarily unsympathetic to social welfare and safety nets, but after 1947 the tone shifted, and neoliberalism's adherents in America (particularly Chicago and Virginia) adopted a much more strident defense of free markets, market rationality and deregulation. Such ideas could not make much headway among a Depression-chastened post-war generation, which believed liberal free markets were unstable, which knew that unemployment could remain stubbornly high for a long time and that government action could bring relief. This generation had, moreover, witnessed the astonishing ability of a centrally directed war economy to mobilize production on a massive scale, as well as the planning achievements of state-led post-war recoveries of devastated nations in Europe and Japan. In this 'collectivist' wilderness, however, the neoliberal faithful did not merely bide their time. They worked steadily and industriously to make important converts where they could, and to spread their gospel via think-tanks like the American Enterprise Institute, the Institute for Economic Affairs, the Heritage Foundation, the Hoover Institution and the Cato Institute.

Friedman particularly was a long-sighted and patient man who believed that, provided he kept chipping away at public, academic and political opinion with his economic theories, his time would come and that it would most likely come in the shape of severe crisis, real or perceived—and when it came people would turn to whatever ideas were 'lying around' (Friedman 1962: xiv; Blyth 2013: 103). He was convinced that if the groundwork were sufficiently laid, it would be his ideas that they found. He was himself a gifted and dedicated simplifier and communicator of neoliberal ideas (and his own revived monetarism) across a transatlantic network (Stedman Jones 2012: Chapter 4).

The ideas of various prominent thinkers usually grouped under the banner of neoliberalism were certainly not uniform, but there was sufficient common ground in their general enthusiasm for policies like deregulation, privatization, anti-unionism, 'letting the market rule', a focus on containing inflation through tight monetary policy rather than on sustaining employment,[17] globalization through trade agreements and opening up developing countries to foreign investment. Daniel Stedman Jones (2012) has demonstrated in some detail that the ultimate triumph of such policies was never a foregone conclusion, and that the adoption of neoliberal ideas was more piecemeal and complex a process than is generally portrayed in images of a Reaganite–Thatcherite revolution. Nevertheless, years of work and proselytization had prepared neoliberals to take advantage of governmental loss of confidence in Keynesian policies in the 1970s, and they had gained intellectual adherents among people with influence on governments. In fact, as Stedman Jones shows, the drift toward neoliberal solutions had already begun under the Democratic administration of Jimmy Carter in the US and the Labour government of James Callaghan in Britain before the clarion call was loudly trumpeted by later conservatives (Stedman Jones 2012: 216–18).

The greatest triumph of neoliberal doctrine, in fact, followed the era of Reagan and Thatcher, and came with the conversion of New Democrats in the US and New Labourites in Britain to neoliberal policies (an enthusiasm later transmitted to labor parties in Australia and New Zealand, and even a socialist party in Germany). Keynes in the midst of world depression, and competition between fascism and communism, had sought a middle way between unfettered capitalism and outright socialism, but the new liberals and laborites proclaimed a 'third way' in politics that was, in fact, an eager capitulation to free market capitalism. The principal beneficiaries were transnational corporations, now able to move comprehensively to offshore production to reduce input costs, a trend given a huge boost after the US Congress establishment of Permanent Normal Trading Relations with China in 2000 made that country a fully legitimate destination for foreign direct investment. Corporate efforts were meanwhile aided and abetted by international institutions like the IMF, World Bank and WTO under the influence of a 'Washington consensus' on neoliberal values. In this mood the lessons of the Depression era were progressively discarded, its strategy of regulatory control believed superseded by sophisticated methods of private sector risk management.[18] Democratic control over and management of the neoliberal economy was believed no longer necessary. Indeed, the beneficent power of free markets to produce steady economic expansion became an article of ideological faith, permitting leaders to argue that citizen welfare required removing impedimentary controls.

The collapse of Soviet communism in 1989 seemed a final vindication of free market policies, yet the actual progress of neoliberalism was marred by setbacks and anomalies. A 1980s 'currency war' following the floating of the dollar led to a series of negotiated revaluations that induced a bubble economy in Japan whose collapse led to long-term stagnation. The adjustment to capitalism in formerly communist countries proved wildly uneven. The 1990s brought a series of rolling financial crises in developing nations under the influence of the Washington consensus, one of the results being the virtual collapse of sub-Saharan Africa. Meanwhile, countries largely insulated from neoliberal shibboleths began to rise, most notably China which joined the capitalist market order but under illiberal Communist Party control.

In the meantime, financial deregulation in the US proceeded apace, with New Deal regulations argued to be obsolete in modern competitive international finance (Hendrickson 2001). Yet the Fed had to increasingly use its lender-of-last-resort power to rescue or reconfigure insolvent financial institutions, while government was forced to organize bailouts of ailing industries after stock market crashes in 1987 and 1989, and a recession in the 1990s. The 'New Economy' of the Clinton era, which envisaged a high-tech, service-oriented US sector complementing a now globalized production sector (Kelly 1999), produced a boom fuelled by unprecedented levels of debt spending by businesses and households, and ended in the dot-com bust and recession of 2000.

These bumps along the way were effectively dismissed by governments and their regulatory agencies because they seemed surmountable using suitable monetary adjustments. What was notable, however, was that recoveries had become 'jobless recoveries' (Groshen and Potter 2003; Schlein 2014). This was a serious issue for the United States because US consumers were, under New Economy assumptions, expected to provide the effective demand that would keep the whole international system going. But with manufacturing jobs siphoned off overseas and middle-class incomes stagnating or declining in real terms over a long period (Leonhardt and Quealy 2014), this function could be performed only through policies of cheap lending and rising debt to maintain consumption levels. One result was a persistent trade deficit as domestic demand sucked in cheap foreign, particularly Chinese, imports (under-priced by virtue of exchange rate suppression), the dollar earnings from which were recycled into US bonds to fuel the easy credit regime run by Alan Greenspan at the Federal Reserve. This regime ended the recession of the early 2000s (another jobless recovery) and then engineered a boom in housing, causing rising prices that made average home-owners feel wealthy, however artificially. So long as the neoliberal economy could be periodically reinflated in this manner, political leaders could feel they were meeting democratic demands, even as the globalization of industry and labor saw domestic employment contract and middle class incomes stagnate.

Having in effect liberated and dispersed the energies of transnational corporations and withdrawn protection from national workers, leaders had inaugurated an interesting experiment to see whether democratic demands could still be fulfilled (or if not fulfilled at least safely contained) within a competitive, globalized economy whose intrinsic political (as opposed to purely economic) aims were never clearly specified. The dedication to preserving employment having been undermined, the only way to fulfill the old social contract's democratic clause was through the expansion of credit, giving birth to a model of global production and consumption that seemed unsustainable in the long run.[19] The illusion was dispelled, finally, when the housing bubble fed a swelling market in mortgage-backed securities among under-regulated financial institutions in pursuit of higher returns, with disastrous consequences for the US, but ultimately also for Europe through the mechanism of accumulated bad bank debt.

And with the onset of financial crisis and prolonged stagnation, the conundrum of the political leader's role in 'economic management', largely obscured by the boom, was cruelly exposed. Leaders were suddenly uncomfortably squeezed as they tried to plot a course between powerful economic and political actors, and a disgruntled public that felt its demands during tough times were being unfairly relegated or ignored. The leadership challenge was to reassert control over wayward economic forces, and especially financial ones, in such a way as to make them once more serve the people rather than dominate them. But where was the new theoretical consensus upon which leaders could converge to both explain the crisis and navigate their way out of it toward securer foundations for growth? It did not exist.

## A failure of consensus

If what Thomas Palley (2012) called the 'flawed paradigm' of neoliberal globalization was now exhausted, as he argued, and an era of prolonged stagnation was in view, a transformation of theory and practice might have been expected. And indeed it was by many, but they were disappointed by the pattern of merely incremental change that emerged. The G8 was expanded into the G20 and the Financial Stabilization Forum yielded to a more formal Financial Stabilization Board, but neither institutional conversion led to a fundamental transformation. Meanwhile, macroeconomic policies were marked by a greater tolerance of inflation and an obsession with correcting fiscal deficits, even as investment and employment languished.

When the neoliberal model of global development cracked in 2007, there was scarcely a leadership cohort anywhere in the world with the confidence or imagination to conceive of an alternate project or, if it could, of how to begin establishing it. There could be no thought, after the collapse of the Soviet Union and the woeful performance of planned economies in general, of a move to socialism. Even the mixed social democratic alternatives that had thrived in European countries in the post-war period had become subject to self-doubt under the increasing strain of trying to meet their social commitments while resisting the pressures of neoliberal competition and neoliberal EU policies. Elsewhere, the neoliberal model had been so thoroughly embraced and implanted that it seemed virtually impossible for political leaders to think outside or beyond it (Mirowski 2013). At most they believed they might patch it up sufficiently with a few renegotiated financial rules and the application of liberal doses of printed currency to get the great credit machine puffing into action once more, which it stubbornly refused to do.

This lack of what may be thought an appropriately dramatic response is less surprising when one considers the lack of any politically available theoretical alternative upon which leaders could converge for collective guidance, such as had existed in the Keynesian moment, and again with the neoliberals in the 1970s and 1980s. Neoclassical economists were naturally not much help and unfortunately their analytical perspective was shared even by Left-leaning neoclassicists like Paul Krugman and Joseph Stigler, who argued strongly for extensive stimulus policies against Right-leaning economists demanding austerity. Such policy recommendations were vitiated by a lack of strong alternative theoretical foundations, and their advocates were easily dismissed as unreconstructed soft core Keynesians whose credentials had long been discredited.

As noted, the heterodox economists had a range of strongly held alternate positions but none had managed to construct, as the neoliberals had done, a robust political network capable of extending beyond adherents to capture the contemporary agenda at a time of crisis. Their failure was perhaps due less to their own theoretical divisions than to the fact that the political audience for their message had shrunk to the point of disappearance. The reconciliation or balancing of liberal economy and democratic polity that is the democratic leader's crucial task had been virtually abandoned during the neoliberal era for the sake of utter reliance on an increasingly liberalized economy that might, so the gamble was, automatically produce the necessary democratic dividend. With the rise of New Democrats, New Labour and a host of neoliberal bureaucrats in the European Union who accepted the logic of this movement, anything resembling even a moderate Left in politics vanished. Any reconstruction now of a scheme of managed capitalism would require, as well as a powerful intellectual justification, a strong political will capable of overcoming the embarrassment of withdrawing from previously declared positions. Such a will with regard to controlling banks may be seen flickering in politicians such as Elizabeth Warren in the US and certain leading figures in the Bank of England, but has hardly been observed in anything like a politically coherent movement anywhere.[20]

## Conclusion: the next crisis?

However much the world may have changed over the course of the last 200 years, we can still descry today the lineaments of the original challenge that industrial capitalism presented to forces of democratization it had itself helped unleash: how, if at all, are these two great movements to be reconciled? It is sometimes forgotten that Karl Marx was necessarily an advocate of entrepreneurial capitalism as much as its adversary. He recognized capitalism as the greatest human productive force ever created and saw it as laying the foundations for a future egalitarian socialist-communist society that could survive above a level of mere subsistence. He argued in effect (from his Hegelian-teleological perspective) that modern enhanced productivity could be combined with democratic justice only over time, with entrepreneurial capitalism a necessary but temporary stage to be ultimately overcome. Those who rejected the Marxist vision, or who were understandably disillusioned by historical attempts to realize it, and who wished to secure the existence and benefits of capitalist development over the long term, had to devise their own political solutions to reconciling economic and democratic imperatives. As Marx would have said, the grand question for the long term is 'Who is to be master, the economy or humanity?'

This, I have argued, is the question posed to all political leaders who must perforce serve the people by serving the economic system that sustains them. They must manage economies to ensure both the continuance of capitalistic growth and the appropriate distribution of its benefits, which means managing the inevitable tensions that arise between liberal economy and democratic polity. The issue is, in other words, simultaneously one of democratic justice and of the stability of global capitalism itself. The current general failure to meet these dual demands is reflected in both popular and academic talk of a 'crisis of leadership' internationally. But leadership failure, I have argued, concerns expert as much as political authority. Leaders must necessarily appeal to an authority outside their own—namely, the authority of those who profess to understand and have mastered the technicalities of a modern economy—if they are to succeed in their political management.

If, as heterodox economists assert, economics cannot be realistically conducted in the absence of considerations of power and distribution, then all economics is automatically political economics, whether it admits it or not, and the connection between political and expert authority is a peculiarly intimate and intricate one in this case. Whether this intricacy affects other social sciences (as opposed to experts in physical sciences) is a moot question, but certainly one can hardly enunciate an economic theory that does not automatically carry political implications. This does not mean, of course, that there is no distinction between expert and political authority. The latter derives from accepted political processes and depends on no necessary possession of expertise at all (in a democratic system); expert authority, on the other hand, is still ultimately founded on the possession of hard, acquired disciplinary skills.

I have argued, however, that expert economic authority is also established through political action, persuasion and statesmanship as much as through economic theory. The vacuum of such authority left by neoliberal experts, whose policies produced an out-of-control financial system and deeply skewed income patterns, has not yet been filled by an alternate, generally recognized economic authority around which a new political consensus can form.[21] Nor does such a consensus seem in immediate prospect. Any attempt to form one, furthermore, will doubtlessly be stoutly contested by stalwart defenders of the current system, however crippled.

Friedman was right in regarding crisis as an opportunity to be grasped. The Obama administration came to office stating the same thing, though signally failed to grasp it on behalf of significant reform. Like other leadership groups, it has given the impression of muddling through with inadequate resources and imperfect understanding. The world now faces the possibility of long-term stagnation, formerly but no longer presumed to be a peculiarly Japanese malaise. How long the global economy can stumble along in this manner is anyone's guess, but if deep structural flaws in the system remain unaddressed, the world risks another, greater crisis that will surely challenge to the limit the ability of governments to manage, but which may also concentrate the minds of politicians and economists more intensely and with more practical effect than we have recently seen.

## Notes

1 No one was surprised when, in the wake of the Global Financial Crisis (GFC) of 2008, incumbent governments in Britain, Spain, Greece, France, Ireland, Holland, Portugal, Belgium and Iceland were judged culpable and thrown from office.
2 This was what Karl Polanyi (2001 [1944]) called the necessary 'double movement' by governments in a fully marketized economy: on one hand, freedom had to be extended to businesses to experiment, change and even fail; on the other hand, mass society had to be protected against the pernicious effects of relentless accumulation. Excessive identification with either of these movements defines, of course, the traditional Right and Left of modern politics, with 'centrists' seeking balance through the 'mixed economy'. Polanyi's point was that, whatever one's political predilections, it belongs to the logic of a system that separates economic from social and political relations that failure to achieve a stable balance puts the whole system at peril.
3 Giving workers a solid and increasing material stake in the system was the key to diluting the revolutionary force that Marx and his followers had expected to be generated by the 'immiseration of the proletariat'. John Rawls's *A Theory of Justice* (1971) can be properly interpreted as a post-hoc philosophical justification of this same solution (unequal distribution being justified only if it advantages the least well off). It was an historical irony that his book appeared just as economic fortune turned against it.
4 A term bestowed to indicate a reduction in the volatility of business cycles due allegedly to increased reliance on the monetary policies of independent central banks rather than on government's fiscal policies (Bernanke 2004). In the policy confusion following the 2008 crash, the monetary policy of central banks, buying back government debt or toxic assets to inject billions of new money into the banking system, emerged once more as the de facto only game in town.
5 Léon Walras, the 'father' of general equilibrium theory, relentlessly pursued mathematization under idealized conditions explicitly in order to align economics with the physical sciences The idealized conditions were: individual utility maximization; the transparency of the qualities of all goods; contracts covering future contingencies are always honored; no one can distort market prices; and everyone is paid at their true productive value. The real world of markets represents an array of 'imperfect' departures from the ideal (Walras 1954).
6 Though, of course, rule of law, in liberal theory, is aimed precisely at restricting the exercise of governmental powers within a framework of formal laws.
7 The neoclassicists used Smith's account to make the self-interested maximization of preferences the foundation of economic thought (Smith 1970). Much work has been done to retrieve a more complex Smith from his interpreters (see Winch 1978; Medema 2009).
8 It is instructive to compare this view with Rousseau's very different conclusions about how market forces corrupt and alienate natural desires to appreciate the ideological import of either view (Rousseau 1973). Smith's version formed the basis of a defense of capitalism, Rousseau's the foundation for capitalism's Marxist critics. Yet it is interesting to note that each tradition has a problematical relationship with the political realm, or more specifically, with 'the state'. In a sense, Marx had a more positive view of the state—the committee of the bourgeoisie governing in favor of the capitalist class—than liberal economists who tended to see it less as facilitating and more as destructively interfering. The economy is fundamental in both but seen as perennially vulnerable to the state in liberal theory and as ultimately triumphant over it in Marxism. Each tradition devalues the political in different ways—liberalism because the state threatens the 'natural' workings of the economy, Marxism because the state was expected inevitably to bend and break before the force of economic necessity.

9 It was a wish realized during the 1970s and 1980s when democratic governments, under pressure of double-digit inflation, removed monetary policy from the purview of politicians by granting statutory discretion to central bankers to act as lender-of-last-resort in crises and to manage interest rates according to strictly 'non-political' criteria, and in so doing put it under the alternate influence of the same financial sector central banks were responsible for regulating (Whitehead 2010).

10 See generally this Special Issue of *Erkenntnis* edited by Grüne-Yanoff (2009) for essays arguing either position.

11 Models, mathematical and otherwise, are constructed on the basis of theories so as to resemble complex reality, always with necessary simplifications that limit the degree of resemblance possible. It is sometimes said that the test of a model lies in its utility rather than its confirmation, which is partly true, but a model that proves its usefulness in testing must go some way toward confirming the accuracy of its foundational assumptions; see Musgrave (1981).

12 It is worth noting that critics of neoclassicism similarly conflate models and theory. John Kay, for example, says that models are metaphors and that the theories informing them are neither true nor false, but rather 'illuminating'. He writes:

> I am arguing that the way we should think of models and use models in economics is to accept many different models. We ought to have a toolkit. The skill of the economist is in deciding which of many incommensurable models he should apply in a particular context.
>
> (Kay 2010: 94–5; see also Bronk 2010)

But what are the factors that make a particular theory 'illuminating', and what are the implicit criteria that enable the economist to exercise the necessary skill in discriminating? John Maynard Keynes seemed to give license for this catholic attitude toward models when he claimed that the skill of choosing was in fact extremely rare. 'Economics', he wrote, 'is a science of thinking in terms of models joined to the art of choosing models which are relevant to the contemporary world. It is compelled to be this, because, unlike the typical natural science, the material to which it is applied is, in too many respects, not homogeneous through time' (cited in Skidelsky and Wigström 2010: 9). Yet even if an economic theory is not eternally applicable (as, say, a theory of gravity) but tied to particular social and historical conditions, it will nevertheless try to say something true about actions and events under those conditions. Keynes's own revolutionary theory involved a rejection of the assumptions underlying equilibrium theory and the substitution of others that seemed to him more accurately descriptive of then current realities (and he labeled it precisely a *general* theory; Keynes 1936).

13 See Fama (1970) who outlined weak, semi-strong and strong versions. The skeptical voices of dissenting 'behavioral' economists were restricted to academia (Shiller 2005; Thaler 2005).

14 The EMH is thus the original source of the old joke about someone strolling with an economist and seeing a $20 bill on the footpath. The economist denies the bill's existence because if it had been there someone would have picked it up already.

15 One of the most pertinent critiques of neoclassical theory, given the nature of the crisis, is its habitual contempt for the money factor (money being regarded as merely a 'veil' over real trade transactions) and thus its neglect of credit-issuing financial institutions in its models. Krugman confirmed the comprehensive neoclassical neglect of credit in a (co-authored) essay of 2010, which, while ostensibly addressing the deficit, constructed yet another model in which neither banks nor money play a causal role (Eggertsson and Krugman 2010). For an analysis of the neoclassical failings in this matter and an argument about the need to bring finance and financial institutions back in, see Boyer (2013).

16 It is always tempting to interpret 'orthodox' and 'heterodox' in terms of religious faith, which may not be entirely inappropriate in the context of economic theory; see, for example, Nelson (2001).

17 An emphasis given policy legitimacy following Fed Chairman Paul Volcker's stringent attack on inflation using high interest rates from 1979 to 1981.

18 Consider the words of Fed Chairman Ben Bernanke in 2006: 'banking organizations of all sizes have made substantial strides over the past two decades in their ability to measure and manage risks' (Bernanke 2006). It was a classic example of the belief that 'this time is different' (Reinhart and Rogoff 2009)—until it isn't.

19 Ignoring the problem of excessive credit was encouraged by the neoclassical view that credit is always someone else's debt and therefore cancels out over the whole. The contrary view presumes that banks, in response to entrepreneurial demand, produce an *increase* in spending power beyond the existing commodity base by simultaneously creating money and debt, causing real macroeconomic effects (see Schumpeter 1934: 101). In post-Keynesian terms, money is *endogenous* to the system, expanding and contracting with the needs of production in response to expectations of aggregate demand, through the banking system (Arestis and Eichner 1988).

20 The Syriza victory in Greece showed the will, but the party's radical promises could scarcely be fulfilled while Greece remained within the straitjacket of the euro, which the party's leader, Alexis Tsipras, refused to abandon, leaving his only option begging Germany for (another) bailout or debt to be written down.

21 The process whereby this happened has been labeled 'financialization', meaning that financial markets, financial institutions and financial elites gain predominant influence over economic policy and economic outcomes, with generally harmful effects. Indicative is the fact that between 1970 and 2010 the ratio of bank assets to GDP among 14 advanced economies grew from about 70 percent to over 200 percent, meaning that banks' assets were now more than twice the value of all goods and services produced in these 14 big economies (Admati and Hellwig 2014).

## References

Admati, A. and Hellwig, M. 2014. *The Bankers' New Clothes: What's Wrong with Banking and What to Do about It*. Princeton, NJ: Princeton University Press.

Arestis, P. and Eichner, A.S. 1988. 'The Post-Keynesian and Institutionalist Theory of Money and Credit.' *Journal of Economic Issues*, 22(4): 1003–21.

Becker, G.S. 1976. *The Economic Approach to Human Behavior*. Chicago: University of Chicago Press.

Bernanke, B.S. 2004. 'The Great Moderation.' Remarks to the Eastern Economic Association. Washington DC, February 20. Retrieved at: www.federalreserve.gov/BOARDDOCS/SPEECHES/2004/20040220/default.htm

Bernanke, B.S. 2006. 'Modern Risk Management and Banking Supervision.' Address to the Stonier Graduate School of Banking. Washington, DC, June12. Retrieved at: www.federalreserve.gov/newsevents/speech/Bernanke20060612a.htm

Blyth, M. 2013. *Austerity: The History of a Dangerous Idea*. Oxford, Oxford University Press.

Boyer, R. 2013. 'Macroeconomics After the Crisis: Bringing Finance Back In.' In *Beyond the Global Economic Crisis: Economics and Politics for a Post-Crisis Settlement*. M. Benner (ed.). Cheltenham: Edward Elgar, 72–93.

Bronk, R. 2010. 'Models and Metaphors.' In *The Economic Crisis and the State of Economics*. R. Skidelsky and C.W. Wigström (eds). New York: Palgrave Macmillan, 101–10.

Dymski, G.A. 2014. 'The Neoclassical Sink and the Heterodox Spiral: Political Divides and Lines of Communication in Economics.' *Review of Keynesian Economics*, 2(1): 1–19.

*The Economist*. 2009. 'Efficiency and Beyond.' July 16. Retrieved at: www.economist.com/node/14030296

Eggertsson, G.B. and Krugman, P. 2010. 'Debt, Deleveraging, and the Liquidity Trap: A Fisher-Minsky-Koo Approach.' Authors' working paper, November 16. Retrieved at: www.princeton.edu/~pkrugman/debt_deleveraging_ge_pk.pdf

Fama, E. 1970. 'Efficient Capital Markets: A Review of Theory and Empirical Work.' *Journal of Finance*, 25(2): 382–417.

Friedman, M. 1953. 'The Methodology of Positive Economics.' In *Essays in Positive Economics*. M. Friedman (ed.). Chicago: University of Chicago Press, 3–43.

Friedman, M. 1962. *Capitalism and Freedom*. Chicago: University of Chicago Press.

Greenspan, A. 2013. 'Never Saw It Coming: Why the Financial Crisis Took Economists by Surprise.' *Foreign Affairs*, November/December. Retrieved at: www.foreignaffairs.com/articles/140161/alan-greenspan/never-saw-it-coming

Groshen, E.L. and Potter, S. 2003. 'Has Structural Change Contributed to a Jobless Recovery?' *Current Issues in Economics and Finance*, 9(8). Federal Reserve Bank of New York. Retrieved at: www.newyorkfed.org/research/current_issues/ci9–8.html

Grüne-Yanoff, T. (ed.) 2009. 'Economic Models as Credible Worlds or Isolating Tools?' Special Issue of *Erkenntnis*, 70(1).

Hartwell, R.M. 1995. *A History of the Mont Pèlerin Society*. Indianapolis, IN: Liberty Fund.

Hendrickson, J.M. 2001. 'The Long and Bumpy Road to Glass-Steagall Reform: A Historical and Evolutionary Analysis of Banking Legislation.' *American Journal of Economics and Sociology*, 60: 849–79.

Kane, J. 2014, 'Leadership Judgment in Economic Affairs.' In *Good Democratic Leadership: On Prudence and Judgment in Modern Democracies*. J. Kane and H. Patapan (eds). Oxford: Oxford University Press, 178–97.

Kay, J. 2010. 'Knowledge in Economics.' In *The Economic Crisis and the State of Economics*. R. Skidelsky and C.W. Wigström (eds). New York: Palgrave Macmillan, 91–100.

Kelly, K. 1999. *New Rules for the New Economy*. Harmondsworth: Penguin.

Keynes, J. M. 1936. *The General Theory of Employment, Interest, and Money*. London: Macmillan.

Kingdon, J. 1995. *Agendas, Alternatives, and Public Politics*, 2nd edn. New York: HarperCollins.

Lazear, E.P. 2000. 'Economic Imperialism.' *The Quarterly Journal of Economics*, 115(1): 99–146.

Leonhardt, D. and Quealy, K. 2014. 'The American Middle Class Is No Longer the World's Richest.' *New York Times*, April 22. Retrieved at: www.nytimes.com/2014/04/23/upshot/the-american-middle-class-is-no-longer-the-worlds-richest.html

Lucas Jr, R.E. 1988. 'On the Mechanics of Economic Development.' *Journal of Monetary Economics*, 22: 3–42.

Lucas Jr, R.E. 2009. 'In Defence of the Dismal Science.' *The Economist*, August 6. Retrieved at www.economist.com/node/14165405

Medema, S.G. 2009. *The Hesitant Hand: Taming Self-Interest in the History of Economic Ideas*. Princeton, NJ: Princeton University Press.

Mirowski, P. 2013. *Never Let a Serious Crisis Go to Waste: How Neoliberalism Survived the Financial Meltdown*. New York: Verso.

Mirowski, P. and Plehwe, D. (eds). 2009. *The Road From Mont Pelerin: The Making of the Neoliberal Thought Collective*. Cambridge, MA: Harvard University Press.

Musgrave, A. 1981. '"Unreal Assumptions" in Economic Theory: The F-twist Untwisted.' *Kylos*, 34(3): 377–87.

Nelson, R.H. 2001. *Economics as Religion: From Samuelson to Chicago and Beyond*. Philadelphia, PA: Pennsylvania State University Press.

Palley, T.I. 2012. *From Financial Crisis to Stagnation: The Destruction of Shared Prosperity* and *the Role of Economics*. Cambridge: Cambridge University Press.

Polanyi, K. 2001 [1944]. *The Great Transformation: The Political and Economic Origins of Our Time*. Introduction by F. Block. Boston, MA: Beacon Press.

Rawls, J. 1971. *A Theory of Justice*. Cambridge: Belknap Press.

Reinhart, C.M. and Rogoff, K.S. 2009. *This Time is Different: Eight Centuries of Financial Folly*. Princeton, NJ: Princeton University Press.

Rousseau, J-J. 1973. 'A Discourse on the Origins of Inequality.' In *The Social Contract and Discourses*, trans. G.D.H. Cole. London: Dent.

Schlein, L. 2014. 'Global Economy Risks Jobless Recovery.' *Voice of America*, January 20. Retrieved at: www.voanews.com/content/ilo-weak-global-recovery-has-curtailed-job-growth/1833672.html

Schumpeter, J.A. 1934. *The Theory of Economic Development: An Inquiry into Profits, Capital, Credit, Interest and the Business Cycle*. Cambridge, MA: Harvard University Press.

Shiller, R. 2005. *Irrational Exuberance*, 2nd edn. Princeton, NJ: Princeton University Press.

Skidelsky, R. and Wigström, C.W. (eds). 2010. *The Economic Crisis and the State of Economics*. New York: Palgrave Macmillan.

Slattery, D., Nellis, J., Josifidis, K. and Losonc, A. 2013. 'Neoclassical Economics: Science or Neoliberal Ideology?' *European Journal of Economics and Economic Policies: Intervention*, 10(3): 313–26.

Smith, A. 1970 [1776]. *The Wealth of Nations*. Harmondsworth: Penguin Books.

Stedman Jones, D. 2012. *Masters of the Universe: Hayek, Friedman, and the Birth of Neoliberal Politics*. Princeton, NJ: Princeton University Press.

Thaler, R.H. 2005. *Advances in Behavioral Finance*, vol. 2. Princeton, NJ: Princeton University Press.

Turner, R. 2007. '"The Rebirth of Liberalism": The Origins of Neo-Liberal Ideology.' *Journal of Political Ideologies*, 12(1): 67–83.

Walras, L. 1954 [1874]. *Elements of Pure Economics*. Trans. W. Jaffe. London: Allen & Unwin.

Weber, M. 2004 [1919]. 'Politics as a Vocation.' In *The Vocation Lectures*. D. Owen and T.B. Strong (eds). Trans. R. Livingstone. Indianapolis, IN: Hackett Publishing Company.

Whitehead, L. 2010. 'The Crash of '08.' *Journal of Democracy*, 21(1): 45–56.

Winch, D. 1978. *Adam Smith's Politics: An Essay in Historiographic Revision*. Cambridge: Cambridge University Press.

# 6 Uneasy expertise

## Geoengineering, social science, and democracy in the Anthropocene

*Clare Heyward and Steve Rayner*

There are few instances where the expertise of natural scientists and engineers has been scrutinized and questioned as much as the scientific literature and discourse on climate change. As others have noted (Thompson and Rayner 1998; Hulme 2009; Kahan 2014), the challenges to the scientific findings are most often politically grounded. Experts in the natural and engineering sciences have, with a few exceptions, responded by trying to think of ways to make the general public understand the often complicated facts about climate change. Social science experts, particularly those with a background in public engagement and deliberation, have frequently been recruited to this task. The 'deficit model' of public understanding of science—according to which public scepticism of scientific claims is due to a cognitive failure to understand scientific arguments—has been largely discredited (Irwin and Wynne 1996; Miller 2001). However, many social scientists have seemed content to accept the agenda of the natural and engineering scientists and play the 'handmaiden' role of translating the pronouncements of scientific experts into action to prevent 'dangerous' climate change (e.g. Moser and Dilling 2004, 2007).

Despite these efforts, political action on climate change has been slow enough to cause despair in a section of the scientific community, leading it to advocate research and development of technologies to achieve, in the words of Shepherd *et al.* (2009: 1), 'the deliberate manipulation of planetary systems to counteract anthropogenic climate change'. (For other prominent examples, see Crutzen 2006; Blackstock *et al.* 2009; Long *et al.* 2011; Government Accountability Office 2010; House of Commons 2010; Rickels *et al.* 2011; National Academy of Sciences 2015a, 2015b). This heterogeneous range of potential technologies is often given a single term, *geoengineering*, and is conventionally divided into carbon dioxide removal (CDR) methods and solar radiation management (SRM) methods (e.g. Shepherd *et al.* 2009). Stratospheric Sulphate Aerosols (SSA) is an example of a SRM method and the most often discussed example of a proposed geoengineering technology.

We emphasize that all geoengineering technologies, including SSA, are proposals. No complete geoengineering technology exists yet. There are some pieces of equipment that could be adapted for deployment as part of a geoengineering technology, but we are still far from any kind of socio-technical system capable of achieving geoengineering goals in a controlled fashion. Geoengineering technologies are currently what STS scholars refer to as socio-technical or technological 'imaginaries' (Jasanoff and Kim 2009), where the specific conditions under which they could be developed and deployed remain uncertain and implicit. The factors affecting the social acceptability of geoengineering technologies, such as the resource requirements, engineering techniques, financing and, most importantly for our purposes, governance arrangements that would be needed for any geoengineering system to operate are thus, as yet, relatively unexplored and therefore unelaborated.

Notwithstanding the indeterminate characteristics of whatever geoengineering technologies might emerge, critics have challenged some of them, especially SSA, on the grounds of democratic legitimacy (e.g. ETC Group 2009). While social scientists addressing geoengineering generally recognize that such technologies raise real issues about how they might be effectively and democratically governed, some seem to have already decided that such technologies are inherently or 'constitutionally' undemocratic (e.g. Macnaghten and Szerszynski 2013; Hulme 2014). However, other social scientists in the field argue that it seems too early to make any kind of judgement about technologies as yet unformed (Markusson 2013; Rayner and Heyward 2013; Healey 2014).

It is also noteworthy that while some have raised concerns about the implications of climate change policies and the related idea of the Anthropocene (e.g. Leach 2013), more generally social scientists, including some critics of geoengineering, have been slower to question the legitimacy of government policies and international institutions designed to achieve drastic emissions reductions. Indeed, some social scientists seem to have become more critical of geoengineering technological imaginaries and at a much earlier stage than they have been of proposals for drastic emissions cuts in the mainstream climate change discourse. But calls to use social science to change people's perceptions and priorities to facilitate 'climate friendly' policies and accept limits to economic growth also raise issues about democracy. We highlight this situation because it raises questions about the role of social science expertise.

The early engagement of social scientists in the field of geoengineering is to be welcomed, especially in the light of their long-standing calls for more 'upstream' public engagement in the early stages of the development of new technologies (e.g. Wilsdon and Willis 2004). However, this engagement is revealing some problematic issues with the idea of upstream engagement, not least the implied model of linear development, which assumes that the final form of a technology can be adequately anticipated by extrapolating from early ideas or 'imaginaries'.

We are also concerned that social science expertise may be shaping or configuring the public responses to geoengineering technologies, while claiming to reflect them. We seek to highlight this issue for the social sciences and to give some reasons why social scientists and others seem more ready to seize on potential governance implications of geoengineering technologies than on those raised by the emissions reduction (mitigation) agenda.

We begin with the recent emergence of geoengineering discourses, in which social scientists have played a role from the early stages. We identify some features of complementary discourses of tipping points and the Anthropocene and highlight their technocratic overtones. We proceed to discuss an example of the critical stance being taken by social scientists and then move on to point out the asymmetry in social scientists' attitudes towards geoengineering and more conventional responses to climate change. Finally, we offer an explanation of this asymmetry, drawing on the Cultural Theory of Mary Douglas.

## The emergence of geoengineering discourses

Just as the political issue of anthropogenic climate change emerged from expert scientific discourse, the same is true of the latest development, the proposal that geoengineering should be considered among the various policy responses to climate change. In both cases, earth systems scientists have set the agendas. However, whereas social scientists seemed fairly content to follow the imperatives set by earth systems scientists when it came to implementing mitigation, there has been immediate resistance among some social scientists to the case for geoengineering, particularly for SSA (e.g. Hulme 2012).

The idea of geoengineering is not new. Proposals to address climate change that would now be regarded as geoengineering have been occasionally mooted since the 1960s. Cesare Marchetti (1977) first used the term specifically in relation to climate change, but to refer to what would now be called Carbon Capture and Storage. Climate geoengineering remained very much on the disreputable fringes of the climate discourse for many years, most climate scientists viewing it as a taboo subject (Lawrence 2006; for an exception, see Keith 2000). This initial reluctance remains reflected in the often expressed concerns about what has been termed 'moral hazard' (Shepherd *et al.* 2009; Keith *et al.* 2010)—that even conducting research into geoengineering techniques will undermine the case to reduce greenhouse gas (GHG) emissions among policy makers or the broader public. Paul Crutzen—winner of the Nobel Prize for his work on stratospheric ozone depletion—broke the scientific community's taboo in a paper in *Climatic Change*, arguing for research into stratospheric sulphate aerosols as a potential geoengineering technology (Crutzen 2006). Since then, as noted above, several reports from scientific and science policy bodies have appeared, or been commissioned by governments, along with a flurry of articles arguing for increased efforts and funding in research of geoengineering techniques.

The idea that geoengineering could be required to deal with a future 'climate emergency' was a strong theme in many early articles advocating research efforts into geoengineering. Following Crutzen, many climate scientists argued that there was a pressing need for research into SSA because it promises to be a fast acting, high leverage technology that could be deployed to avert abrupt climate events (see, for example, Blackstock *et al.* 2009; Victor *et al.* 2009; Caldeira and Keith 2010; Blackstock and Long 2011; Long *et al.* 2011; Goldblatt and Watson 2012; Victor *et al.* 2013). While there are other arguments for SSA research, the climate emergency argument was a key argumentative strategy. As has been argued elsewhere (Heyward and Rayner forthcoming 2015), this particular argument arguably broke the scientific community's taboo on the advocacy of geoengineering research.

The climate emergency justification was itself facilitated by a growing emphasis on abrupt climate events and especially 'tipping-point rhetoric' in mainstream climate discourse. Although the idea of sudden discontinuities has its origins in mathematical 'catastrophe theory' (Thom 1975), the term 'tipping point' was popularized by the mainstream author Malcolm Gladwell (2000) to describe sudden discontinuous social change. Russill and Nyssa (2009) describe in detail how it was subsequently imported into the climate discourse by Hans Joachim Schellnhuber and introduced into peer-reviewed scientific papers by James Hansen. What began as a strategy used by scientists to communicate climate change to non-experts became the subject of research papers (e.g. Lenton *et al.* 2008; Kriegler *et al.* 2009; Lenton 2011) and invoked in various research agendas.

Tipping-point rhetoric emphasizes irreversible, abrupt and catastrophic climatic changes. Those who employ it typically argue for immediate measures to avoid crossing an irreversible threshold (e.g. Rockström *et al.* 2009). The introduction of tipping-point rhetoric into mainstream climate discourse thus primed audiences for the idea of a climate emergency. Whereas the original users of tipping-point rhetoric urged immediate curbs on GHG emissions, climate emergency rhetoric opened up a different course of action—namely, research into SSA.

The Anthropocene is another theme that has been invoked in support of geoengineering research. The term was originally coined by Eugene Stoermer and popularized by Paul Crutzen (2002), who suggested that the influence of humankind on fundamental earth systems had become so significant as to usher in a new geological era. He wrote:

> For the past three centuries, the effects of humans on the global environment have escalated. Because of these anthropogenic emissions of carbon dioxide, global climate may depart significantly from natural behaviour for many millennia to come. It seems appropriate to assign the term 'Anthropocene' to the present, in many ways human-dominated, geological epoch, supplementing the Holocene—the warm period of the past 10–12 millennia. The Anthropocene could be said to have started in the latter part of the eighteenth century, when analyses of air trapped in polar ice showed the beginning of growing global concentrations of carbon dioxide and methane. This date also happens to coincide with James Watt's design of the steam engine in 1784.
> (Crutzen 2002: 23)

Others have taken up the concept, which is now the subject of many scientific research articles, conferences, research agendas and at least two scientific journals.[1] As the era of the Anthropocene is increasingly invoked, it is not surprising that different participants employ it in different ways. Depending on the predilections of the speakers, the discourse of the Anthropocene might be intertwined with warnings about tipping points (e.g. Biermann *et al.* 2012) and the need not to exceed planetary boundaries (Rockström *et al.* 2009). Sometimes it is connected directly with the need for geoengineering research (Crutzen 2002: 23; Steffen *et al.* 2007: 619). Regardless of the final prescriptions for action, there are some common features in the discourses of the Anthropocene. The acceptance of these features in general discourse means that the idea of geoengineering the climate no longer seems so alien or unthinkable.

The concept of the Anthropocene asserts that humanity and the planet that it inhabits have entered a new era, but there are at least two different ways in which this era is characterized. One focuses on the overwhelming biophysical effects of human activity, which Crutzen dates as beginning in the eighteenth century. Others argue that it has a much longer history—for example, beginning with the advent of agriculture (Ruddiman 2003). The argument over the date of origin is politically charged, with commentators who are most worried about the prospect of passing a global tipping point favoring the later date. In their view, pushing back the onset of the Anthropocene by several millennia dilutes the concept's immediate mobilizing power.

Another way of characterizing the Anthropocene focuses on human consciousness of its role in shaping the planet rather than the advent of that role. In the words of *New York Times* journalist Andrew Revkin: 'Two billion years ago, cyanobacteria oxygenated the atmosphere and powerfully disrupted life on earth. . . . But they didn't know it. We're the first species that's become a planet-scale influence and is aware of that reality. That's what distinguishes us' (cited in Stromberg 2013). Steffen *et al.* (2007) divide the Anthropocene into three sub-stages: the 'Industrial Era' (circa 1800–1945), the 'Great Acceleration' (circa 1945–2015) and finally, a potential future, 'Stewards of the Earth?' from 2015 onwards. The main feature of the last sub-stage is that it is characterized by '[t]he recognition that human activities are indeed affecting the structure and function of the Earth System as a whole (as opposed to local and regional scale environmental issues) . . . filtering through to decision-making at many levels' (Steffen *et al.* 2007: 618). The new era is effectively one of human consciousness: the recognition that humanity has great effects on the planetary environment. Regardless of when the power of humans became so great (agriculture or the Industrial Revolution), humanity is only now starting to realize this fact.

A further feature of the Anthropocene discourse follows from this recognition. Having realized the power of humanity over the planetary environment, it is argued that we must change our ways of thinking and our ways of acting. Steffen *et al.* (2007) express the hope that humankind will, over the next few years, wake up to the impacts that it is having on global systems and do whatever it takes, by adopting new technologies, changing values and behavior or, most likely, a combination of both, to ensure that life can continue. The idea of the Anthropocene is thus invoked to do political work.

Moreover, as currently expressed, the political agenda of the Anthropocene has strong technocratic overtones. We are told repeatedly that expert scientific understandings must affect political decisions and change human behavior. The discourse of the Anthropocene might be regarded as the latest instance of a long-standing human propensity to appeal to nature to justify moral and political preferences. For example, Rayner and Heyward (2013) describe how at the 2012 'Planet under Pressure' conference, 'reality' and 'nature' were frequently invoked by natural scientists as the impetus for political action. Johan Rockström, the lead author of the influential 'planetary boundaries' hypothesis, drove home the point claiming 'We are the first generation to know we are truly putting the future of civilization at risk'. Social scientists were taken to task for failing in the handmaiden role. Dutch political scientist Frank Biermann claimed that 'The Anthropocene requires new thinking' and 'The Anthropocene requires new lifestyles'—to be ushered in by the application of social science expertise (cited in Rayner and Heyward 2013: 141). The *State of the Planet Declaration*, produced by the Planet Under Pressure conference, stated that 'consensus is growing that we have driven the planet into a new epoch, the Anthropocene' and called for a 'new contract between science and society in recognition that *science must inform policy to make more wise and timely decisions*' (Brito and Stafford-Smith 2012: 6, our emphasis). The Anthropocene is thus taken to mandate the greater involvement of scientific experts in policy making. Crutzen himself wrote, 'A daunting task lies ahead for *scientists and engineers* to guide society towards environmentally sustainable *management* during the age of the Anthropocene' (Crutzen 2002, our emphasis). We are told that scientific expertise is needed to tell policy makers when planetary boundaries are in danger of being exceeded, tipping points approached and to guide policy making to ensure that disaster is averted and that critical planetary systems are maintained.

Like tipping-point rhetoric, the Anthropocene discourse offers the possibility of at least two quite different technocratic futures relying on somewhat different forms of technical expertise. One emphasizes fundamental behavioral and institutional transformation to ensure that humanity develops within 'planetary boundaries' set by earth systems scientists. In this framework, the expertise of social scientists is to be harnessed to persuade politicians and publics of the need to act and to design policy instruments to achieve the goals determined by the scientific experts. The other vision, less sanguine about the prospects of changing embedded socio-technical practices and overcoming vested interests, would deploy engineering expertise to cut the Gordian knot of inexorably rising greenhouse gas emissions through geoengineering.

Both potential futures highlight the idea that new ways of thinking are needed: in the case of the second vision, priming audiences to think about whether it is now permissible to intervene in planetary systems at a large scale. After all, the success of any geoengineering technology is predicated on the idea that nature can (and now should) be carefully managed. For both visions, the prospect of control over the Earth's biological, chemical and physical systems marks the completion of the transition to the third stage of the Anthropocene, 'wise stewardship' of nature. The whole planet becomes a garden, to be managed by those who have the requisite skills and experience. In both futures, management is dominated by scientific expertise. Both raise challenges for democratic governance. Social scientists have noted such tensions. For example, Melissa Leach recently couched the issue in the following terms:

> Is there a contradiction between the world of the Anthropocene, and democracy? The Anthropocene, with its associated concepts of planetary boundaries and 'hard' environmental threats and limits, encourage[s] a focus on clear single goals and solutions . . . It is co-constructed with ideas of scientific authority and incontrovertible evidence.
>
> (Leach 2013)

Leach does not contest the idea that there are 'natural limits' or non-negotiable targets for sustainable development. Rather, she points out that the role given to scientific experts in identifying them could mean that the same experts have a more than appropriate role in determining the pathways and courses of action that should properly be left to contestation of interests and values characteristic of democratic politics.

Both the geoengineering vision and the planetary boundaries imaginary may be equally dependent on natural science and engineering expertise to determine just how much should be done, when and where. But the geoengineering discourse seems to propose a subtler role for the social sciences than is the case with the planetary boundaries version of the Anthropocene. Whereas the latter largely looks to social science expertise to persuade the public to implement its agenda, geoengineering discourses have, thus far, overwhelmingly looked to social science expertise to explore the social conditions under which even researching, let alone implementing, such measures would be widely viewed as even permissible, let alone desirable (e.g. Bunzl 2009; Shepherd *et al.* 2009; Keith 2013; Hulme 2014). Of course, it can be argued that both visions of the Anthropocene ultimately cast the social sciences in a service role and that the geoengineering imaginary is simply at an earlier stage of engaging with them. But, at the present moment, the different approaches to the deployment of social science expertise seem real and relevant as geoengineering researchers insist on the necessity of broad public consultation and engagement in the governance of research into geoengineering technology (e.g. Shepherd *et al.* 2009; SRMGI 2011; Carr *et al.* 2013). The motivations behind such calls are many and mixed, ranging from a largely pragmatic instrumental concern about securing public acceptance or a social licence to operate, through the idea that public input will result in a substantively more robust technology, to the conviction that it would simply be wrong to proceed without prior informed consent from the public.

However, some social scientists have suggested that we can already answer the question of the social acceptability of geoengineering. Phil Macnaghten and Bronislaw Szerszynski (2013) go beyond the concerns that we have already expressed about the technocratic impulses associated with the idea of the Anthropocene to advance the claim that there is already public recognition that at least some forms of geoengineering, specifically SSA, cannot plausibly be realized in ways that would be compatible with democracy. There are two intertwined claims here. One is that there is, indeed, something inherently undemocratic about SSA. The other is that such concerns reflect public, rather than expert, concerns. In the following section we engage with both arguments because of their substantive importance and because of the issues that they exemplify for the exercise of social science expertise in the context of geoengineering and other emerging technologies.

## Geoengineering and democracy

In the conclusions of a paper reporting focus group research into public perceptions of SSA, Macnaghten and Szerszynski (2013: 472) claim that the technology has 'an anti-democratic constitution' that is incompatible with liberal democracy. In support of this conclusion they identify four themes arising from their focus group discussions:

1 Conditional acceptance of SSA.
2 Scepticism of climate science as a reliable guide to policy or as able to predict side-effects, and whether the technologies could be tested at a sub-deployment scale, both leading to the concern that human beings will be the guinea pigs in a climate experiment.
3 Concern that technology would become politicized and used in ways that are radically at odds with intended purpose of countering climate change.
4 Lack of confidence in capacity of existing political systems to accommodate SSA.

They conclude that the public in their focus groups revealed a 'more consistently sceptical position about the prospect of geoengineering than has been reported in earlier research' and question 'whether solar radiation management can be accommodated within democratic institutions, given its centralizing and autocratic "social constitution"' (Macnaghten and Szerszynski 2013: 472). Points 3 and 4 clearly raise issues of legitimacy and thus relate to their worry that there is significant potential for SSA to 'negate democracy'.

Clearly, there are many grounds to be concerned whether existing systems can manage SSA. We share these concerns ourselves and have sought to address them for geoengineering generally in a work of explicit advocacy (Rayner and Heyward 2013). But two questions arise. First is whether these concerns are sufficient to support the claim that SSA necessarily has a centralizing and autocratic social constitution—and that, therefore, it is inherently or inevitably undemocratic. Second is whether these concerns genuinely arise from the public's own understanding of the technology or whether they were perhaps informed, even stimulated, by the researchers' own expert interpretations of the issues rooted in their understanding of and experience with past technological controversies.

To address the first question of whether SSA is constitutionally undemocratic, we need to unpack two ideas. One is the model of technological development underlying the notion that a technology, in this case SSA, has a 'constitution' that is already established at the early stage where it is merely a socio-technical imaginary. Second is the question: what counts as 'democracy'?

The idea that a technology has a social constitution derives from the work of Langdon Winner (1980), who refers to technologies as 'inherently political artifacts' in both a stronger and a weaker sense. The stronger sense refers to those that require or entail specific social arrangements. The weaker sense refers to those that are strongly compatible with a particular sociological system. He further distinguishes conditions internal to the workings of a given technical system and those that are external to it. Macnaghten and Szerszynski's reference to the 'social constitution' of SSA would seem to suggest both the strong and internal sense. But even the weak and external sense implies that the eventual form that SSA technology may take can already be specified with sufficient precision to reach the conclusion that it is constitutionally incompatible with a democratic context. In turn, this suggests a linear model of technology development whereby the speculative projections of natural science and engineering experts at an early stage of the development of technological artifacts are sufficient for social science expertise to be able predict their political entailments (Markusson and Wong 2015).

The idea that social scientists can already specify the future social constitution of a technology that barely exists beyond the imagination of a few engineers and natural scientists is at odds with long-standing critiques of linear models of technological development (e.g. Godin 2006). Linear models give insufficient weight to uncertainty and contingency in suggesting an orderly progression from scientific conception, through research, to technological completion, underplaying the scope for human choice and 'user' agency in favor of a predictable technical outcome (Rip 1995). Just as engineering advocates of new technologies are likely to project highly optimistic visions of a technology's future (Linden and Fenn 2003; Rayner 2004), contemporary social scientists may be inclined to emphasize potentially negative outcomes. Under the best of circumstances, the interplay of technology promoters and detractors can lead to a process of distributed technology assessment or social learning (Rayner 2004). Under less felicitous conditions, it can lead to social conflicts and misallocation of societal resources. Social science expertise can legitimately play two roles in such processes. One is as observer, commentator or, even, facilitator of the process of 'distributed technology assessment'—the role that Roger Pielke (2007) describes as the 'honest broker'. Another is as a social critic of a specific technological proposal—the role Pielke describes as 'advocacy'. In this instance, Macnaghten and Szerszynski seem to have conflated these two functions of social science expertise. To explain our concern we turn first to their definition of democracy.

Macnaghten and Szerszynski explicitly base their conviction that SSA is constitutionally undemocratic on the claim that its effects are manifest on a planetary scale and that it must be controlled centrally, presumably by a global body. They write:

> Democracy, in its various forms, depends on the articulation, negotiation and accommodation of plural views and interests. It relies on an evolving and partially flexible relationship between citizens and governance institutions. Solar radiation management [SSA] *by contrast exists as a planetary*

> *technology*. While plausibly able to accommodate diverse views into the formulation of its use, once deployed, *there remains little opportunity for opt-out or for the accommodation of diverse perspectives*. By its social constitution it appears inimical to the accommodation of difference. *Following deployment it could only be controlled centrally and on a planetary scale.*
>
> (Macnaghten and Szerszynski 2013: 472, our emphasis)

Thus, Macnaghten and Szerszynski's view of democracy requires that individuals not only have a right to express dissent, but have either the ability to *determine*, rather than influence, whether SSA is used, or the ability to 'opt out'—to be able to live in an environment free of the use and effects of SSA. This draws on a particular model of democracy that has been common to much STS scholarship since the 1970s and is rooted in the ideal of a tight-knit community of highly engaged citizens, where intense deliberation is possible (Barber 1984; Lengwiler 2008; Durant 2011). Such a model of democracy favors small, decentralized units of consensual decision-making and is generally suspicious of political mechanisms for large political units, which distance executive power from citizens. From this point of view, aggregative or representative forms of democracy (Lovbrand *et al.* 2011) are seen as inferior or even stalking horses for authoritarianism.

The consensual participatory conception of democracy can appear rather demanding. If being able to opt out of decisions were a standard feature of any political institution, the result would be more akin to anarchy than democracy. For example, ordinary citizens do not have a general right to opt out of their governments' laws, even if they disagree with them. The fact that we cannot refuse to pay taxes, decide to drive on the wrong side of the road, or carry a machete on to the London Underground does not mean by itself that the UK is an undemocratic country (even if there are other reasons for asserting that it is). Thus, on a reasonable view of democracy, living in a political community means some restrictions on behavior, including restrictions on the right to opt out of certain decisions. From this point of view, what makes a political community democratic is whether people have an adequate say in the decisions that affect their lives. It is not to guarantee that the decisions will go their way, or that they will be subject only to decisions that they have actually consented to. Nor is it enough merely to assume or assert that global or centralized political structures must be undemocratic. A system where any institution governing SSA would be part of a multilateral global order, subject to checks and balances, is at least conceptually possible. Implementation would, of course, require much political work and diplomatic negotiation and perhaps it would turn out to be impracticable. The historical record of global institutions is mixed and progress uneven. But we cannot simply write off the evolution of international law and institutions since the Second World War. We do not claim that multilateral governance of SSA would necessarily be democratic, but equally we would challenge the notion that it must necessarily be anti-democratic. We recognize the potentially undemocratic implications of technologies that are powerful enough to affect the global climate. However, as we have argued, it seems at least premature to impose inherent characteristics on an unformed socio-technical imaginary.

Returning to our concern that Macnaghten and Szerszynski conflated two types of social science expertise, we address the question of how the concerns about democracy arose within the focus groups. Macnaghten and Szerszynski acknowledge that these concerns did not appear spontaneously. Rather, they began by introducing the subject of SSA under the conventional frame of the perceived need to buy more time for greenhouse gas mitigation policies to become effective, but subsequently introduced additional possible framings. These were perspectives from environmental and civil society actors and the geopolitical history of weather and climate modification, for which James Fleming's (2010) book, highly critical of geoengineering, was the only reference given.[2] Thus, the concerns about democratic governance were raised by the focus group participants only after the introduction of more critical materials highlighting 'the possible use of solar radiation management techniques or social, political, and military purposes unrelated to climate change policy' (Macnaghten and Szerszynski 2013: 468).

This methodological move suggests that, rather than emerging spontaneously from the focus group, responses may have been configured by the authors' own concerns about SSA and its incompatibility with their particular model of democracy (the advocacy role). While Macnaghten and Szerszynski concede that 'some may argue that [our framing] may have unduly shaped public responses' (2013: 472), they argue that they were attempting to 'open up' the debate (Stirling 2008) by the use of deliberative methods (the honest broker role). However, there remains a real question of whether this is rather a case of what Pielke (2007) describes as 'stealth advocacy'. There seem to be grounds for concern that the researchers configured their public in accordance with their own critical image of geoengineering as 'anti-democratic'. Such a move would, of course, have the effect of enhancing the apparent legitimacy of this view by attributing it to a broader public than expert social scientists. To be clear, we are not suggesting that Macnaghten and Szerszynski were deliberately seeking to co-opt the public. However, their paper does seem to highlight the challenges facing social scientists seeking to 'open up' issues in the 'honest broker' role at such an early stage of technology development, where the future form of that technology and the society in which it might be realized remain as indeterminate as SSA.

It seems that, in common with most members of the natural sciences community (including some of those who actively pursue research and development of geoengineering technologies) and many environmental activists and campaigners, social scientists are deeply troubled by the prospect of using certain geoengineering technologies. This raises a potential problem for those engaged in social science research on geoengineering. Should the social science expertise be deployed to shape what the public thinks, or focus on identifying pre-existing public opinion? Does focusing on the anti-democratic potential of SSA give a misleading impression to the public that problems of global governance and legitimacy are intrinsic to SSA, or more broadly to geoengineering, but are not equally worrisome when it comes to the extensive social engineering called for in relation to emissions mitigation measures?[3] Hence, we return to our earlier observation that concerns about democracy have come to the forefront of social science engagement with the idea of a geoengineered Anthropocene sooner and more forcefully than has been the case with the planetary boundaries version.

## Geoengineering and social engineering: a curious asymmetry

We noted earlier an apparent asymmetry in the social science attention to the implications of geoengineering for democracy compared to mitigation and other conventional climate change responses. If global institutions and planetary management systems were, per se, threats to democracy, then one would expect similar concerns to arise with respect to proposals for 'Earth Systems Governance' and management of 'planetary boundaries' as proposed by Biermann *et al.* (2012). These include radical institutional reforms in the name of effective Earth System Governance comprising: the 'upgrading' of the UN Environmental Programme so that it becomes a specialist UN agency with a sizeable role in agenda setting, norm development, compliance management, scientific assessment and capacity building (the environmental equivalent to the World Health Organization); measures to further integrate sustainable development policies at all levels; and closing gaps in global regulation, especially of emerging technologies. These proposals for global institutions and stringent policies designed to achieve radical changes in human behavior to 'save the planet' would seem to be at least as distant from Macnaghten and Szerszynski's ideal of grassroots democracy as any that would be required to govern SSA.

Perhaps it is simply the case that the problematic implications for democracy are being made far more visible much earlier in the case of SSA. We have noted that early climate emergency justifications for SSA certainly exhibited a worrisome coercive quality. Research, and research *now*, into SSA was presented as the only possible course of action that could avert a possible climate emergency (e.g. Caldeira and Keith 2010).[4] Moreover, 'emergency' arguments are discomfiting. Declarations of a 'state of emergency' have been used many times by political leaders both in autocratic and largely democratic countries to justify oppressive political action and close down dissent and debate. This feature of emergency rhetoric was noted, seemingly approvingly, in an early report on geoengineering: 'in a crisis, ideological objections to solar radiation management may be swept aside' (Lane *et al.* 2007: 12). By using such bold statements and 'state of [the planet] emergency arguments', we might say that the early proponents of SSA research effectively invited scrutiny of the distribution of political power and the possibility of authoritarianism.

However, the idea that we have only a short time to save the planet has long been deployed in support of drastic climate policies and the planetary boundaries imaginary of the Anthropocene, without attracting the extent and intensity of concern about democracy that SSA seems to evoke. The explicit deployment of the climatic emergency argument may partially account for the greater concern with democratic control of the response strategy that we see with regard to geoengineering. It seems insufficient explanation on its own.

Here we turn to political anthropology for a complementary perspective. Mary Douglas's Cultural Theory (Douglas 1970, 1978; Thompson *et al.* 1990; Thompson and Rayner 1998) points out that people's expectations, views and preferences are linked to the different social contexts in which they find themselves throughout their lives. What is visible, obvious or even blatant in one social context is ignored, overlooked or invisible in another (Rayner 2012). It is not enough merely to assert that early proponents of SSA made obvious (perhaps unwittingly) the potential governance problems whereas those advocating mitigation and adaptation did not. The question to be answered is why the anti-democratic line of critique was almost instantly highlighted in the case of geo-engineering but has remained muted throughout the much longer history of conventional mitigation and adaptation policies.

Cultural Theory posits three active 'voices' in public debates, each offering different stories about human nature, individual-group relations, the natural world, risk, responsibility and distributive justice. The voices are termed *egalitarian*, *individualist* and *hierarchical*. Politics, according to Cultural Theory, can be understood in terms of fluctuations in the relative power of these three politically active voices. Often alliances are made between two of the voices against the third, but they are rarely permanent.

The egalitarian voice is committed to universal participatory democracy in which every citizen has an equal voice and decisions are made through consensus, not unlike the model seemingly invoked by Macnaghten and Szerszynski. It also holds that nature is very delicately balanced and the slightest perturbations can lead to catastrophe. Hence, the egalitarian prescription in environmental politics is to respect nature's fragility and make only minimal demands—the quintessential Green 'tread lightly' approach. Concern for the vulnerable, natural or human, is a driving value for the egalitarian voice. It has spoken loudly and clearly in debates on climate change through green NGOs and activist scientists. For example, in the words of Pene Lefale (1995):

> What is causing corals to die lies at the core of the way we humans live. . . . Dead corals are the victims of injustices . . . of greed, of selfishness. . . . It is an act of genocide. . . . The coral polyp's own world mirrors the human experience—the cries for freedom from foreign debt, poverty, starvation, the cries to change lifestyles, not the climate, the cries to stop burning fossil fuels! To ignore the death of coral reefs is, I believe, to ignore the cries of many of the world's people.

The hierarchical perspective is also community-oriented but, unlike the egalitarian, does not require active participation and explicit consent from all of its citizens to legitimate every political decision. Day-to-day decisions are best left to expert managers applying well-established rules and procedures. Ensuring stability is best achieved by strengthening formal national and intergovernmental institutions, such as the UN, as advocated by Biermann (supra). In the words of the UN Human Development Report (UNDP 1999: 2), 'The challenge is to find the rules and institutions for stronger governance'. This includes:

- A stronger and more coherent UN system . . .
- A global central bank
- A world investment trust . . .
- A revised World Trade Organization . . .
- An international criminal court . . .
- A broadened United Nations . . .

(UNDP 1992: 110–11)

The hierarchical voice describes nature as having limits and how transgressing them can lead to catastrophe. However, it is less certain than the egalitarian that the world is on the edge of catastrophe and is perennially seeking to determine exactly where the limits are through expert assessments, such as those of the Intergovernmental Panel on Climate Change or the Millennium Ecosystems Assessment. Maintaining the social and natural order is a driving value here. However, from the hierarchical viewpoint, natural systems are not always on the verge of collapse and natural systems can be exploited to some degree and managed with suitable skill.

The individualist voice emanates, not from collective organization, but from loose networks of people connected by individual market-style transactions, rather than webs of mutual obligation. Accordingly, decisions are the aggregate of individuals' own private decisions or the outcomes of competition rather than participatory deliberation or the operation of due process. The individualist voice trusts that markets will yield optimal outcomes for society and distrusts interference in their operation by governments. The individualist view of nature describes it as robust and able to withstand any human interference or exploitation. Hence, it has little patience with prophecies of doom such as those emanating from environmental activists. The individualist driving values are self-reliance and innovation. In global environmental politics, the individualist perspective is most likely to take a sceptical view of climate change to begin with, but also holds that, should it turn out to be a problem, market-driven technological innovation can be trusted to provide solutions. In the words of Newt Gingrich, 'Geo-engineering holds forth the promise of addressing global warming concerns for just a few billion dollars a year. We would have an option to address global warming by rewarding scientific innovation. Bring on American ingenuity. Stop the green pig' (cited by Vidal 2011).

Hence, conventional political action on global climate change can be largely understood as a coalition between the other two voices. The egalitarian voice, expecting catastrophe, appeals to hierarchical values of maintaining order resulting in a political consensus for emissions mitigation. In this way, egalitarians are able to harness hierarchical means to serve their ends of curtailing the individualist propensity towards excessive and inequitable consumption. However, in order to do this, 'the egalitarian must moderate their suspicion of authority' (Thompson *et al.* 1990: 89). In getting their calls for mitigation heeded, the egalitarian has had to overlook the 'uncomfortable knowledge' (Rayner 2012) of the hierarchical version of democracy and legitimacy, and its tendencies towards technocracy and possible authoritarianism. The egalitarian–hierarchical call for mitigation of greenhouse gases was, for many years, regarded as the only acceptable course of action in the face of anthropogenic climate change. Even adaptation was once regarded as taboo (Pielke *et al.* 2007) and the 'fragile planet' image of the egalitarian worldview, with support from hierarchical views, became hegemonic in climate change discourse (Rayner 1995).

SSA has raised the prospect of another route by which hierarchy could achieve system stability because, unlike the egalitarian, the hierarchical viewpoint is not intrinsically committed to the 'tread lightly' approach to nature. To date, hierarchy has accepted mitigation and, more recently, adaptation as the only available forms of responsible climate change management. The emergence of the idea of geoengineering, especially SSA, challenges this position. Geoengineering, we might say, is the ultimate in environmental management, in terms of its scale and complexity. As a form of environmental management, it is congruent with the hierarchical perspective. As a proposal for more, rather than less, intervention in natural systems, it is in stark opposition to the egalitarian perspective. It also opens up the possibility for hierarchy to shift from alliance with egalitarianism to one with competitive individualism, based on the promise that SSA could be implemented at significantly lower cost per unit of avoided temperature increase than the cost of conventional mitigation. The prospect of lower costs appeals to both the system maintenance value associated with hierarchy, as well as to competitive individualism that sees SSA as a relatively low cost form of insurance in the event that their climate scepticism proves misplaced (see, for example, Levitt and Dubner 2009).

If the hierarchical and egalitarian perspectives start to disagree about the appropriate responses to climate change—raising the prospect of a thought coalition with the individualist voice—then there is less incentive for the egalitarian voice to withhold criticism of 'undemocratic' hierarchical means. We suggest that the intense early concerns about the implications of SSA for democracy can be understood as reflecting concerns that it will weaken this long-standing alliance in environmental politics between hierarchical and egalitarian approaches, perhaps leading hierarchy to lean in the direction of the individualist voice on climate change and forging a new coalition in which the egalitarian voice carries less weight. From the egalitarian standpoint SSA does, indeed, represent a threat to its conception of democracy in that it would lead to a weakening of its voice in the political sphere. However, we should be careful not to equate democracy with any of the three voices. Each has its own account of democracy (Wildavsky 1987) and all are inevitably flawed from the other perspectives. We would argue that democracy *is* the conversation among the three voices (Verweij and Thompson 2006). From the standpoint of Cultural Theory, SSA proposals should be seen not as a threat to democracy, but as a stimulus to democratic discourse.

## Conclusion

The recent emergence of a geoengineering discourse exhibits a complex set of relationships between technoscientific and social scientific expertise. Scientists involved in geoengineering discourse convey mixed messages about the need for technocratic management of the Anthropocene at the same time as expressing strong commitments to the importance of public participation in decision-making about geoengineering.

We have described one case in which social scientists conducting such public engagement experiments report public concern about the compatibility of SSA with democratic decision-making. However, this emerged only after a briefing in which the researchers' own concerns had been introduced, thus raising questions about the relationship between, what we believe, are two legitimate roles for social science that should not be conflated.

First is the 'honest broker' role of opening up discourse, in which the social scientist draws on his or her expertise to ensure that all parties to a debate have the opportunity to be heard and responded to by each other. Sometimes this can legitimately involve amplifying the voices of those who have difficulty being heard. However, social scientists in this role must be careful not to configure those voices to conform to their own views and preferences or allow their own views to be presented as those of their research subjects—a position that Pielke describes as 'stealth advocacy'.

The second exercise of social science expertise is when researchers voice their own concerns, based on their expert analysis of an issue. In such instances, audiences can reasonably expect to see the data and/or expert reasoning behind such informed open advocacy. We disagree with Macnaghten and Szerszynski both with respect to the weight that they place on the ability to opt out in defining democracy and on the issue of whether SSA can be said to already have a socio-technical constitution. However, we have no problem with their presenting these views as expert judgements. We do worry that presenting them in the context of a public engagement exercise obscures the distinction between the expert roles of honest broker and issue advocate.

We have also wondered why SSA has already attracted such early concern about democracy and technocracy, while sometimes quite draconian climate mitigation proposals seem to have escaped such critical attention. Our answer is that the prospect of SSA is viewed as a threat to the established climate policy coalition between hierarchical and egalitarian worldviews based on a shared concern with maintaining system stability.

In this alliance, it was expedient for the egalitarian voice to be muted about potentially unequal distributions of political power in return for acceptance, by the hierarchical perspective, of the idea that the planet upon which humans live is complex and fragile, and that humans' consumption of the planet's resources must be therefore reduced. However, the idea of geoengineering implies that planet Earth is complex but nevertheless manageable by humans. This is compatible with the hierarchical conception of nature, but is an anathema to the egalitarian viewpoint. Therefore, as geoengineering is increasingly discussed, there is less incentive for the egalitarian voice to withhold its criticisms of technocracy and centralized authority. Accordingly, we may expect to see more animated social science criticism of global institutions and scientific power when geoengineering is discussed than when similar political changes and institutions were discussed in pre-geoengineering mainstream climate discourse.

## Notes

1 There is an Elsevier journal entitled *Anthropocene* and a Sage journal *The Anthropocene Review*.
2 Whether the problems associated with previous technological innovations, such as weather modification or biotechnology, can be applied simply to the debates about geoengineering is another area in which some social scientists are insufficiently critical.
3 An equally problematic tendency is to present concerns specific to SSA as problems of 'geoengineering', when that category encompasses a diverse set of imaginary technologies, including other SRM methods and CDR technologies.
4 Some may interpret this argument slightly differently: that Caldeira and Keith do not argue that SSA research is the only way to avoid an emergency, but that research should be started now *if* later a climate emergency does manifest. They may thus see these scientists as making a far more conditional claim and not predicting that a climate emergency will happen. While it is true that Caldeira and Keith say that a climate emergency is possibility, not a certainty, the article leaves the reader in little doubt that research into SSA is a necessity anyway. SSA is presented as the *only* course of action that could avert a climate emergency (should one happen). Therefore, if we are concerned about climate emergencies (which we should be, as there is at least some risk) we should engage in SSA research—and start now.

## References

Barber, B. 1984. *Strong Democracy: Participatory Politics for a New Age*. Berkeley, CA: University of California Press.

Biermann, F., Abbott, K., Andresen, S. *et al.* 2012. 'Navigating the Anthropocene: Improving Earth System Governance.' *Science*, 335(6074): 1306–07.

Blackstock, J.J. and Long, J.C.S. 2011. 'The Politics of Geoengineering.' *Science*, 327(5965): 527.

Blackstock, J.J., Battisti, D.S., Caldeira, K. *et al.* 2009. 'Climate Engineering Responses to Climate Emergencies.' Santa Barbara, CA: Novim.

Brito, L. and Stafford-Smith, M. 2012. 'State of the Planet Declaration.' Planet Under Pressure: New Knowledge Towards Solutions Conference. London, March 26–9. Retrieved at: www.planetunderpressure2012.net/pdf/state_of_planet_declaration.pdf

Bunzl, M. 2009. 'Researching Geoengineering: Should Not or Could Not?' *Environmental Research Letters*, 4(4): 045104.

Caldeira, K. and Keith, D. 2010. 'The Need for Geoengineering Research.' *Issues in Science and Technology*, 17: 57–62.

Carr, W.A., Preston, C.J., Yung, L., Szerszynski, B., Keith, D.W. and Mercer, A.M. 2013. 'Public Engagement on Solar Radiation Management and Why It Needs to Happen Now.' *Climatic Change*, 121(3): 567–77.

Crutzen, P.J. 2002. 'Geology of Mankind.' *Nature*, 415(6867): 23.

Crutzen, P.J. 2006. 'Albedo Enhancement by Stratospheric Sulfur Injections: A Contribution to Resolve a Policy Dilemma?' *Climatic Change*, 77(3): 211–20.
Douglas, M. 1970. *Natural Symbols*. London: Barrie & Rockcliffe.
Douglas, M. 1978. 'Cultural Bias.' Occasional Paper no. 35. London: Royal Anthropological Institute.
Durant, D. 2011. 'Models of Democracy in Social Studies of Science.' *Social Studies of Science*, 41(5): 691–714.
ETC Group. 2009. 'Geopiracy: The Case against Geoengineering.' Ottowa: ETC Group.
Fleming, J. 2010. *Fixing the Sky: The Checkered History of Weather and Climate Control*. New York: Columbia University Press.
Gladwell, M. 2000. *The Tipping Point: How Little Things Can Make a Big Difference*. London: Little Brown.
Godin, B. 2006. 'The Linear Model of Innovation: The Historical Construction of an Analytical Framework.' *Science, Technology & Human Values*, 31(6): 639–67.
Goldblatt, C. and Watson, A.J. 2012. 'The Runaway Greenhouse: Implications for Future Climate Change, Geoengineering and Planetary Atmospheres.' *Philosophical Transactions of the Royal Society A: Mathematical, Physical and Engineering Sciences* 370(1974): 4197–216.
Government Accountability Office (US). 2010. *Climate Change: A Coordinated Strategy Could Focus Federal Geoengineering Research and Inform Governance Efforts*. Washington, DC: Government Accountability Office, 10–903.
Healey, P. 2014. 'The Stabilisation of Geoengineering: Stabilising the Inherently Unstable?' CGG Working Paper no. 15. Institute for Science, Innovation and Society, University of Oxford.
Heyward, C. and Rayner, S. (forthcoming 2015). 'Apocalypse Nicked! Stolen Rhetoric in Early Geoengineering Advocacy.' In *Anthropology and Climate Change*, 2nd edn. S. Crate and M. Nuttall (eds). Walnut Creek: Left Coast Press.
House of Commons Select Committee on Science and Engineering (UK). 2010. *The Regulation of Geoengineering*. London: The Stationery Office.
Hulme, M. 2009. *Why We Disagree about Climate Change: Understanding Controversy, Inaction and Opportunity*. Cambridge: Cambridge University Press.
Hulme, M. 2012. 'Climate Change: Climate Engineering through Stratospheric Aerosol Injection.' *Progress in Physical Geography*, 36(5): 694–705.
Hulme, M. 2014. *Can Science Fix Climate Change? A Case Against Climate Engineering*. Cambridge: Polity Press.
Irwin, A. and Wynne, B. 1996. *Misunderstanding Science: The Public Reconstruction of Science and Technology*. London: Routledge.
Jasanoff, S. and Kim, S-H. 2009. 'Containing the Atom: Sociotechnical Imaginaries and Nuclear Power in the United States and South Korea.' *Minerva*, 47(2): 119–46.
Kahan, D. 2014. 'Making Climate-Science Communication Evidence-Based: All the Way Down.' In *Culture, Politics and Climate Change*. M. Boykoff and D. Crow (eds). Abingdon: Routledge.
Keith, D.W. 2000. 'Geoengineering the Climate: History and Prospect.' *Annual Review of Energy and Environment*, 25(1): 245–84.
Keith, D.W. 2013. *A Case for Climate Engineering*. Cambridge: MIT Press.
Keith, D.W., Parson, E. and Morgan, G.M. 2010, 'Research on Global Sun Block Needed Now.' *Nature*, 463 (7280): 426–7.
Kriegler, E., Hall, J.W., Held, H., Dawson, R. and Schellnhuber, H.J. 2009. 'Imprecise Probability Assessment of Tipping Points in the Climate System.' *Proceedings of the National Academy of Sciences*, 106(13): 5041–6.

Lane, L., Caldeira, K., Chatfield, R. and Langhoff, S. 2007. 'Workshop Report on Managing Solar Radiation.' NASA, Hannover, MD, April.

Lawrence, M. 2006. 'The Geoengineering Dilemma: To Speak or Not to Speak.' *Climatic Change*, 77(3): 245–8.

Leach, M. 2013. 'Democracy in the Anthropocene? Science and Sustainable Development Goals at the UN.' *Huffington Post*, March 28. Retrieved at: www.huffingtonpost.co.uk/Melissa-Leach/democracy-in-the-anthropocene_b_2966341.html

Lefale, P. 1995. 'Statement by NGOs to the Plenary.' COP 1, Berlin, March 30.

Lengwiler, M. 2008. 'Participatory Approaches in Science and Technology: Historical Origins and Current Practices in Critical Perspective.' *Science, Technology & Human Values*, 33(2): 186–200.

Lenton, T.M. 2011. 'Early Warning of Climate Tipping Points.' *Nature Climate Change*, 1(4): 201–09.

Lenton, T.M., Held, H., Kriegler, E., Hall, J.W., Lucht, W., Rahmstorf, S. and Schellnhuber, H.J. 2008. 'Tipping Elements in the Earth's Climate System.' *Proceedings of the National Academy of Sciences*, 105(6): 1786–93.

Levitt, S.D. and Dubner, S. J. 2009. *Superfreakonomics: Global Cooling, Patriotic Prostitutes and Why Suicide Bombers Should Buy Insurance.* New York: William Morrow.

Linden, A. and Fenn, J. 2003. 'Understanding Gartner's Hype Cycles.' Strategic Analysis Report no. R–20–1971. Gartner, Inc. Stamford, May 30.

Long, J., Raddekmaker, S., Anderson, J. *et al.* 2011. 'Geoengineering: A National Strategic Plan for Research on the Potential Effectiveness, Feasibility and Consequences of Climate Remediation Technologies.' Washington, DC: Bi-Partisan Policy Center.

Lovbrand, E., Pielke Jr, R. and Beck, S. 2011. 'A Democracy Paradox in Studies of Science and Technology.' *Science, Technology and Human Values*, 36(4): 474–96.

Macnaghten, P. and Szerszynski, B. 2013. 'Living the Global Social Experiment: An Analysis of Public Discourse on Solar Radiation Management and Its Implications for Governance.' *Global Environmental Change*, 23(2): 465–74.

Marchetti, C. 1997. 'On Geoengineering and the CO2 Problem.' *Climatic Change*, 1(1): 59–68.

Markusson, N. 2013. 'Tensions in Framings of Geoengineering: Constitutive Diversity and Ambivalence.' CGG Working Paper no. 19. Institute for Science, Innovation and Society, University of Oxford.

Markusson, N. and Wong, P-H. 2015. ''Geoengineering Governance, the Linear Model of Innovation and the Accompanying Geoengineering Approach.' CGG Working Paper no. 20. Institute for Science, Innovation and Society, University of Oxford.

Miller, S. 2001. 'Public Understanding of Science at the Crossroads.' *Public Understanding of Science*, 10(1): 115–20.

Moser, S. and Dilling, L. 2004. 'Making Climate Hot.' *Environment: Science and Policy for Sustainable Development*, 46(10): 32–46.

Moser, S.C. and Dilling, L. (eds). 2007. *Creating a Climate for Change: Communicating Climate Change and Facilitating Social Change.* Cambridge: Cambridge University Press.

National Academy of Sciences. 2015a. *Climate Intervention: Carbon Dioxide Removal and Reliable Sequestration.* Washington, DC: National Academy of Sciences.

National Academy of Sciences. 2015b. *Climate Intervention: Reflecting Sunlight to Cool Earth.* Washington, DC: National Academy of Sciences.

Pielke Jr, R. 2007. *The Honest Broker.* Cambridge: Cambridge University Press.

Pielke Jr, R., Prins, G., Rayner, S. and Sarewitz, D. 2007. 'Lifting the Taboo on Adaptation.' *Nature*, 445(7128): 597–8.

Rayner, S. 1995. 'Governance and the Global Commons.' In *Global Governance*. M. Desai and P. Redfern (eds). London: Pinter, 60–93.

Rayner, S. 2004. 'The Novelty Trap: Why Does Institutional Learning about New Technologies Seem so Difficult?' *Industry and Higher Education*, 18(6): 349–55.

Rayner, S. 2012. 'Uncomfortable Knowledge: The Social Construction of Ignorance in Science and Environmental Policy Discourses.' *Economy and Society*, 41(1): 107–25.

Rayner, S. and Heyward, C. 2013. 'The Inevitability of Nature as a Rhetorical Resource.' In *Anthropology and Nature*. K. Hastrup (ed.). Abingdon: Routledge, 125–46.

Rickels, W., Klepper, G., Dovern, J. *et al.* 2011. 'Large-scale Intentional Interventions into the Climate System? Assessing the Climate Engineering Debate.' Scoping Report conducted on behalf of the German Federal Ministry of Education and Research (BMBF). Kiel: Kiel Earth Institute.

Rip, A. 1995. 'Introduction to New Technology: Making Use of Recent Insights from Sociology and Economics of Technology.' *Technology Analysis and Strategic Management*, 7(4):417–32.

Rockström, J., Steffen, W., Noone, K. *et al.* 2009. 'A Safe Operating Space for Humanity.' *Nature*, 461(7263): 472–5.

Ruddiman, W.F. 2003. 'The Anthropogenic Greenhouse Era Began Thousands of Years Ago.' *Climatic Change*, 61(3): 261–93.

Russill, C. and Nyssa, Z. 2009. 'The Tipping Point Trend in Climate Change Communication.' *Global Environmental Change*, 19(3): 336–44.

Shepherd, J., Caldeira, K., Cox, P. *et al.* 2009. 'Geoengineering the Climate: Science, Governance and Uncertainty.' London: The Royal Society.

SRMGI (Solar Radiation Management Governance Initiative). 2011. 'Solar Radiation Management: The Governance of Research.' London: The Royal Society.

Steffen, W., Crutzen, P.J. and McNeill, J.R. 2007. 'The Anthropocene: Are Humans Now Overwhelming the Great Forces of Nature.' *AMBIO: A Journal of the Human Environment*, 36(8): 614–21.

Stirling, A. 2008. '"Opening Up" and "Closing Down": Power, Participation, and Pluralism in the Social Appraisal of Technology.' *Science, Technology & Human Values*, 33(2): 262–94.

Stromberg, J. 2013. 'What is the Anthropocene and Are We in It.' *Smithsonian Magazine*, January. Retrieved at: www.smithsonianmag.com/science-nature/what-is-the-anthro pocene-and-are-we-in-it-164801414/?no-ist.

Thom, R. 1975. *Structural Stability and Morphogenesis*. French edn [1972]. Trans. D.H. Fowler. London: Benjamin.

Thompson, M., and Rayner, S. 1998. 'Cultural Discourses.' In *Human Choice and Climate Change: An International Assessment, Volume I, The Societal Framework*. S. Rayner and E.L. Malone (eds). Columbus: Battelle Press, 256–343.

Thompson, M., Ellis, R. and Wildavsky, A. 1990. *Cultural Theory*. Boulder, CO: Westview Press.

UNDP (United Nations Development Programme). 1999. 'Human Development Report.' New York: Oxford University Press.

Verweij, M. and Thompson, M.(eds). 2006. *Clumsy Solutions for a Complex World: Governance, Politics and Plural Perceptions*. Basingstoke: Palgrave.

Victor, D., Morgan, M.G., Apt, J., Steinbruner, J. and Ricke, K. 2009. 'The Geoengineering Option: A Last Resort against Global Warming?' *Foreign Affairs*, 88(2): 64–76.

Victor, D. and Morgan, M.G., Apt, J., Steinbruner, J. and Ricke, K. 2013. 'The Truth about Geoengineering: Science Fiction and Science Fact.' *Foreign Affairs*, March 27.

Retrieved at: www.foreignaffairs.com/articles/139084/david-g-victor-m-granger-morgan-jay-apt-john-steinbruner-kathari/the-truth-about-geoengineering.

Vidal, J. 2011. 'Geoengineering: Green Versus Greed in the Race to Cool the Planet.' *The Guardian*, July 10. Retrieved at: www.theguardian.com/environment/2011/jul/10/geo-engineering-weather-manipulation

Wildavsky, A. 1987. 'Choosing Preferences by Constructing Institutions: A Cultural Theory of Preference Formation.' *American Political Science Review*, 81(1): 3–22.

Wilsdon, J. and Willis, R. 2004. *See-Through Science: Why Public Engagement Needs to Move Upstream*. London: Demos.

Winner, L. 1980. 'Do Artifacts Have Politics?' *Daedalus*, 109(1): 121–36.

# 7 Democratic and expert authority in public and environmental health policy

*David Kriebel and Daniel Sarewitz*

'Without chemicals', the old Monsanto ad explained, 'life itself would be impossible'. The implications of this truism become ever more complex as a profusion of new chemicals are introduced into the world's industrial enterprises each year, making their way into modern lives through products and processes, jobs and economic value, as well as through ecological and human-biological pathways. Industrial chemicals are a building block of modernity, and specific chemicals and combinations of chemicals are also a threat to the health and welfare of humans and ecosystems. Balancing the value of chemicals for modern livelihoods and ways of life with the threat to human and environmental wellbeing is a complex undertaking negotiated at the intersection of science, politics and policy. In the United States, nearly 40 years after a chemical regulatory regime, the *Toxic Substances Control Act* (TSCA), was passed into law, the outcome of this negotiation is extremely clear: more than 84,000 chemicals have been inventoried by the US Environmental Protection Agency (EPA), and just five[1] have been regulated under the TSCA to limit their production (USEPA 2012a; CRS 2008).

If one eliminates from the inventory polymers (which are generally thought to have minimal adverse effects from toxicity) and chemicals produced at levels of less than 10,000 pounds per year, the number drops to about 15,000 (CRS 2008). Yet no one suggests that even this smaller number is well characterized in terms of environmental and human health effects, and even for the approximately 3,000 chemicals produced at more than one million pounds (approximately 454 metric tons) annually, 'basic information about chemical properties is lacking' (CRS 2008). EPA itself argues that TSCA needs to undergo significant reform 'to improve EPA's chemical management authorities' (USEPA 2012a). Yet it is also the case that the core operational assumption behind the existing legal framework, and any proposed reforms, remains essentially unchanged after 40 years: the foundation of chemical regulation must be scientific information. This logic is explicitly set out in the language of TSCA legislation itself, whose initial finding is that 'adequate data should be developed with respect to the effect of chemical substances and mixtures on health and the environment'. EPA's authority to regulate derives from the existence or creation of this data. As its first principle for TSCA reform, 40 years later EPA asserts that '[s]ound science should be the basis for the assessment of chemical risks, while recognizing the need to assess and manage risk in the face of uncertainty' (USEPA 2012b).

While the idea that scientific data must be the foundation for regulatory decisions seems almost trivially obvious, this intuition rests on a particular understanding of science that is in many ways not applicable to highly contested and uncertain regulatory environments. EPA, and others promoting TSCA reform, understand the lack of progress in regulating toxic chemicals as a problem of EPA's legal authority both to compel chemical manufacturers to provide necessary scientific information, and to promulgate and enforce regulatory decisions that will hold up to judicial scrutiny (Jacobs *et al.* 2010; Belliveau 2011). In other words, administrative and legal authority are understood to be the main obstacles to effective regulatory action. Yet the democratic legitimacy of this authority derives from a deeper source, that epistemic authority delegated to science itself, which is deemed unproblematic so long as it is available and 'sound'—that is, the idea that scientific findings correspond to actual truths in the world provides the underlying (if unstated because it is supposedly obvious) basis for science-based regulatory and legal authority. However, this separation of scientific and legal authority is incoherent. In a key 1991 ruling, the US Court of Appeals overturned an EPA effort to regulate asbestos partly on a judgement that the EPA had overstepped its regulatory authority, but also on the basis of EPA's failure to compare the risks of asbestos to the risks of alternative materials, and its failure to adequately assess the costs of the ban[2]. The judicial decision thus makes clear that there is no obvious boundary that separates the science of risk from the politics of risk and the values necessarily associated with various risk framings, such as fairness (e.g. Rayner and Cantor 1987). This point is made still more apparent by the fact that, through amendments to the TSCA, Congress has acted directly several times to provide for regulation of specific chemicals and materials in specific settings, including asbestos in schools (amendments of 1986 and 1988), indoor radon (amendment of 1988) and residential lead paint (amendment of 1992), thus pre-empting exactly the science-based regulatory process that TSCA was created to pursue (Schierow 2013). Meanwhile, of course, the regulated community—that is, the chemicals manufacturing industry—has continually done whatever it could to prevent the EPA from taking action (and to avoid having to provide it with the scientific information upon which such action could be justified), and its legal victory in the asbestos ruling made very clear the hurdles that the EPA would have to clear to strongly regulate any chemical under TSCA (Cannon and Warner 2011).

The *Occupational Safety and Health Act* (OSHAct) of 1970 is another bulwark of the federal government's laws governing toxic chemicals and their health effects for workers. Permissible Exposure Limits (PELs) are among the most important aspects of the law, specifying 'safe' levels in workplace air. In the first year of the law, almost 450 PELs were defined. For chemicals that were either in widespread use or particularly hazardous, OSHA established not only a PEL but also a standard—a detailed set of specific rules governing safe handling of the chemical. But this process almost immediately became the subject of protracted legal battles, and in the first nine years of the agency, OSHA managed to issue fewer than two standards per year (Jacobs *et al.* 2010).

One of the chemicals for which a PEL was issued in the first round in 1971 was methylene chloride, a widely used organic solvent (Jacobs *et al.* 2010). As with many of the initial set of PELs, OSHA adopted a 'consensus standard' from the American Conference of Government Industrial Hygienists (ACGIH), a professional organization. Since 1946, ACGIH had recommended that methylene chloride in factory air be kept below 500 parts per million (ppm) to protect against neurologic and cardiovascular effects, and in 1971, OSHA adopted this number. Then in 1985 the National Toxicology Program (NTP), the government's chief laboratory for toxicity testing, found that methylene chloride was carcinogenic in mice and rats. This new evidence strongly suggested that the chemical was more hazardous than previously thought and that the old standard of 500 ppm was much too high. Using a special provision of the OSHAct, the United Autoworkers Union petitioned OSHA to issue an emergency temporary standard to protect workers from cancer risk, while a permanent standard was developed. About a year and a half later, the petition was denied. In 1997, 12 years later, OSHA finally issued a rule lowering the permissible exposure limit from 500 ppm to 25 ppm (OSHA 1997).

The cases of asbestos and methylene chloride are typical—the 'one chemical at a time' approach to regulation has many such examples. Indeed, many of the toxics recognized as major environmental health hazards decades or even centuries ago are still heavily used and continue to kill people despite overwhelming evidence of their danger: benzene, asbestos, and lead are three prime examples (Kriebel 2009).

## Toxic chemicals and science in the real world

Underlying these failures to act to prevent disease from toxics are two interrelated problems. First, there is a lack of clarity on the part of both scientists and policy makers about how much evidence of risk is sufficient to justify any particular line of preventive actions. This uncertainty leads to confusion and delay, creating fertile ground for the second problem: economic actors and their agents who use the contested terrain of the science to promote their interests at the expense of competing interests motivated mostly by non-economic motives, especially protection of human and environmental health. Thus, the two problems are linked: uncertainty about how much evidence is enough to justify action is created by the existence of competing interests, some favoring action, others not. Put another way, the consequent debates are a manifestation of unacknowledged limits of science. The question of whether a body of scientific information is sufficient or 'sound' enough to justify a regulatory decision is an essentially political one.

The problem of sufficient knowledge and certainty is particularly difficult because the human body is a complex system whose dynamics remain in many important respects poorly understood. As one example, the specific cellular events leading to lung cancer from tobacco smoke are known only in broad outline, despite billions of dollars of cancer research. To make this point more stark: when a smoker dies of lung cancer, no existing medical tests—not even at post-mortem examination—can tell whether the tumor that killed him was caused by tobacco carcinogens, or something else. We may be able to say that it is *likely* that he died of tobacco-related cancer, but we cannot be certain. For weaker toxic effects than that of tobacco, the problem is even more challenging. Scientific uncertainty lingers even in the most rigorous study because of unknown factors: seldom do we have strong reason to believe that our scientific models are complete and that no important factors or interactions have been left out (Heij 1989; Hansson 2002; Kriebel 2009). Add to this ignorance about causal mechanisms, our inability to conduct experiments on humans to test toxicity and the challenges of extrapolating from mice to people and we must acknowledge that decisions about how to protect the public from toxics will always be made in the face of uncertainty.

Several decades of social science research on science and risk have demonstrated that science as practiced often plays out as negotiations among experts with differing disciplinary, institutional, and political perspectives to arrive at 'science-based' policy decisions (e.g. Schwarz and Thompson 1990; Sarewitz 2004; Pielke 2007). A valuable framework for understanding this process is Funtowicz and Ravetz's (1990) notion of 'post-normal' science—science that is conducted when facts are uncertain, social values are in dispute, the stakes are high, and decisions are urgent. What is particularly useful—and often inadequately appreciated—about the idea of post-normal science is that it quite literally recognizes and defines a different type of science, one distinct from all the standard definitions and demarcations of conventional, or normal, science. When politicians, scientists, or interested parties in disputes over the regulation of chemicals talk about the need to base decisions on good data or 'sound science' or other such bromides, they are appealing to an ideal of science as practiced in the laboratory under highly controlled conditions and within well accepted theoretical frameworks—normal science, in Kuhn's (1962) terminology. That is not what the science behind regulating chemicals is like. As just one clear indication of how different post-normal science is from conventional ideals of science, just consider that, as Guston (2006) has pointed out, scientists routinely resort to voting in order to settle issues of regulatory science. More generally, Jasanoff (1990) has documented the complex negotiations that scientists undertake among themselves as part of the science advisory process related to the implementation of regulatory regimes.

The US regulatory and scientific response to environmental degradation, public health hazards, and scientific uncertainty over the past 25 years has focused heavily on the development of quantitative risk assessment methods. This approach is portrayed and presumably thought of by its proponents as a normal science in the sense that additional scientific results are supposed to incrementally strengthen and expand existing knowledge frameworks. Such science is not oriented towards seeking solutions, but to creating a higher level of certainty upon which action can then be based. Moreover, the assumption that toxicology, epidemiology, and risk assessment are normal sciences (i.e. science that through controlled experimentation and well-calibrated models provides quantitative data on exposure and toxicity in order to produce accurate estimates of risk to health or the environment) creates a de facto presumption of harmlessness—the more uncertain the risk, the less likely it is to be deserving of rapid and aggressive protective action. There are simply far too many chemicals with poorly understood health effects already on the market for any reasonable expectation that normal science can adequately measure the risk of each one—or even that the single chemical, acting by itself, is the right unit of assessment. A more stringent application of the current regulatory approach—in which the government identifies hazards one at a time and develops regulations for each chemical that firms then adhere to—cannot be effective against the complexity of synthetic materials in our environment.

The limited usefulness of the current normal science approach is illustrated by the problem of chemicals with endocrine disrupting properties. Exogenous disruptors of endocrine function have a very wide range of effects, which vary from species to species, often do not follow simple linear dose-response relationships, and are highly sensitive to windows of vulnerability during development. This creates the condition Stirling (2007) calls *ambiguity*—the inability to confidently identify the adverse outcomes whose risks need to be estimated. In this condition, quantitative risk assessment cannot provide meaningful evidence for choosing among alternatives, yet it continues to be represented as the 'sound science' approach to making decisions on toxic chemicals (Kriebel *et al.* 2001; Stirling 2007).

Conventional notions of science envision a rational process practiced in isolation from, and then applied to, society. This fits the idealized (but, as we have emphasized, not the actual) model of how chemicals come to be regulated under TSCA or the OSHAct. However, the failure of the TSCA to achieve much beyond scientific and regulatory gridlock provides strong empirical support for the idea that the science involved in regulating chemicals is post-normal—if this were not the case, then disputes about chemical regulation would not involve disagreements about science at all, but would simply be played out as explicitly political disputes between parties with conflicting interests and values. In a normal science world, such disputes will always look like this:

> PARTY A: The science says such-and-such, and our belief is that this is sufficient to justify restrictions on production.
>
> PARTY B: Yes, the science does say such-and-such, but our belief is that this is not sufficient to justify restrictions.

In a world of post-normal science, the opposite often pertains. Both sides may claim that they want to protect health and the environment, and to protect jobs and profitability, but they invoke different bodies of evidence, different interpretations of evidence, different views of relevant uncertainties, and so on, in advancing preferences about regulatory actions.

While post-normal science is inappropriate to a linear regulatory regime like TSCA, where science comes first, politics and policy, come second, it offers many other potential alternative pathways for bringing science, policy, and politics together to achieve action. Some of these pathways are now beginning to take shape in the US, characterized by instances of civil society organizations, academic researchers, and private industry working together to limit the production and use of particular chemicals without a federal regulatory mandate and despite what, in normal scientific parlance, would be termed ongoing and in some cases significant scientific uncertainty about the health effects of those chemicals.

## Action without regulation

Formaldehyde is a widely used industrial chemical found in hundreds of household products and building materials. It is also a human carcinogen, according to the National Toxicology Program (NTP 2011). Listing as a carcinogen, however, does not trigger regulatory action in the US, and formaldehyde continues to be legally used in many products, including cosmetics and other personal care products. But Johnson & Johnson, one of the leading manufacturers of such products in the US, began to remove formaldehyde from its personal care products, despite the fact that this is not required by federal law (Thomas 2014). More and more manufacturers and large retailers are taking actions to remove hazardous chemicals from their products and their supply chains without waiting for regulatory action. This surprising trend is partly driven by an awareness in industry that the number of synthetic chemicals of potential concern in consumer products far outpaces the ability of government regulators to evaluate their safety. It is also driven by environmental advocates who see an opportunity to move markets directly, rather than through lobbying and legislation.

Increasingly, manufacturers and retailers realize that waiting for definitive quantitative evidence is a short-term strategy with considerable risk to their economic success. They have seen how costly it can be to wait too long to eliminate a dangerous chemical, incurring legal costs and lost market share through unfavorable consumer perceptions, even if in the end they win the legal battle. Our conversations with large retailers and manufacturers have shown us that many are frustrated with the failure of the 'one chemical at a time' quantitative risk assessment paradigm (as are the regulators and civil society actors) and are looking for different strategies (Rossi *et al.* 2011). Numerous firms, such as Dell Computer, Walmart, and Boots Alliance (a large British health and beauty retailer) have demonstrated a willingness to take preventative action in the face of uncertainty that contrasts markedly with the normal science risk assessment approach currently enshrined in US laws (Kriebel 2001) and thus tacitly, if not explicitly, acknowledging that action need not wait upon incontrovertible evidence of harm (Kriebel 2007).

Environmental advocates have been effective in pushing industries and states to take action to remove toxic chemicals in several targeted sectors (Geiser 2011). They are using market campaigns to replace toxic chemicals in cosmetics, and pushing 'take-back' campaigns to reduce the risk of toxic wastes from discarded electronics. The healthcare sector has been effectively targeted to remove mercury and several other toxics by building a strong industry-labor-advocate coalition. And state legislative campaigns have increasingly been successful; between 2003 and 2010, 71 chemical safety laws were passed in 18 states by overwhelming bipartisan margins (Belliveau 2011).

Three short case studies illustrate a range of different ways that chemical exposures in consumer products have been reduced, in the absence of agreement on the certainty of risk, federal government regulation, and the standard model of normal science legitimated through expert scientific authority.

Bisphenol A (BPA) is an estrogenic chemical used in the manufacture of polycarbonate plastic and epoxy resins. Polycarbonate is a clear, rigid, yet impact-resistant plastic commonly used in sports and baby bottles until recently, and is also used in automobile safety applications, safety helmets, eye glasses, kitchen appliances, CDs and DVDs, and water coolers. It is a relatively expensive polymer valued for its unique properties. Epoxy resins containing BPA are used to coat food and beverage cans and drinking-water filters, and are also used in dental sealants. BPA has been in commercial use since the 1950s and current global production is greater than six billion pounds (approximately 2.7 million metric tons) (Vogel 2009).

A recent CDC study found BPA in the urine of 93 per cent of tested participants, indicating there is widespread human exposure to this chemical (US Centers for Disease Control and Prevention 2009). BPA has been shown to act as a synthetic estrogen, and animal studies have found it to be a developmental, neural, and reproductive toxin. As early as 1997, researchers found adverse impacts on animals at levels below government safety standards (vom Saal *et al.* 1997). In September 2008, the results of the first major epidemiologic study on BPA linked exposure to diabetes and cardiovascular disease (Lang *et al.* 2008). A recent study has found that low-level exposure to BPA in pregnant women may adversely affect the developing fetus (Benachour and Aris 2009). At about the same time, the NTP released a report that raised concern about effects on brain, behavior and prostate gland in fetuses, infants, and children at current human exposure levels (NTP 2008). An accumulating literature since then has strengthened the evidence of hazard (Peretz *et al.* 2014). In 2012, the US Food and Drug Administration amended its regulations to no longer include the use of BPA in baby bottles and children's drinking cups, while continuing to allow other uses—chiefly in food packaging (US Food and Drug Administration 2014). The withdrawal of approval for BPA in baby bottles was triggered by a petition from the chemical industry, which demonstrated that the chemical was no longer in use in these products. In 2013, California placed BPA on the list of chemicals that are considered potentially hazardous and therefore subject to listing on products (called the Proposition 65 list after the citizen initiative that created the requirement).

An EU risk assessment of BPA published in 2008 concluded that products made from BPA are safe for consumers and the environment when used as intended (EC 2008). In 2008, the government of Canada took action on BPA, listing it as a toxic substance and banning the chemical from baby bottles.

Environmental health activists in a number of states—at last count there were 13 including the District of Columbia—have been successful at passing state laws limiting children's exposure to BPA (NCSL 2014).

In 2008, large retailers, including Walmart Canada and Toys R Us, began removing baby products containing BPA following the government of Canada's announcement. Six baby bottle manufacturers agreed to voluntarily remove BPA from baby bottles following a request from attorneys general of Connecticut, Delaware, and New Jersey in 2008 (Rust *et al.* 2009). Nalgene, a maker of sports bottles, announced that it would stop using BPA and recall existing products (Layton and Lee 2008). In March 2009, Sunoco announced that it would sell BPA only to those customers who could guarantee that they would not use the chemical in food and water bottles for children under three (Sissell 2009). Tritan(tm), a new and little-studied co-polyester, is the chemical commonly being substituted for BPA.

In spite of these actions by state governments and corporations and the accumulating evidence of harmful low-dose effects of BPA exposure, some government and academic scientists and the plastics industry continue to argue that BPA is safe in use. The Harvard Center for Risk Analysis (which has received funding from the American Chemistry Council and the plastics industry) released a report in 2004 that gave little weight to the evidence of harm found in low dose studies (Vogel 2009). In addition, the US Food and Drug Administration (FDA) continues to reassure consumers that 'the available information continues to support the safety of BPA for the currently approved uses in food containers and packaging' (US Food and Drug Administration 2014).

Phthalates are a class of industrial chemicals invented in the 1930s and widely used in a variety of consumer products including medical devices, cosmetics, food wrap, building materials, and children's toys made of polyvinyl chloride (PVC). Annual global production of phthalates is estimated to be 11 billion pounds (approximately 5 million metric tons) (McGinn 2000). They are considered one of the most ubiquitous classes of chemicals in the environment. Phthalates are used to soften plastic, as solvents in many applications, and in cosmetics to hold fragrance and make products more effectively penetrate and moisturize the skin. These low-volatility chemicals are not bound in the plastic polymer, so may become bioavailable through migration out of solid objects and condensation on house dust as well as through hand and mouth contact with phthalate-containing products.

In the 1990s, scientific evidence about the potential reproductive health risks of *in utero* and childhood exposure to phthalates began to emerge. In animal studies, these chemicals have been found to damage the liver, kidney, heart, and lung; they also present reproductive and developmental hazards (DiGangi *et al.* 2002). Similarly, there is growing concern about the impacts of phthalates on male reproductive health. A recent study found reduced male-typical play behavior in boys whose mothers had prenatal exposure to anti-androgenic phthalates (Swan *et al.* 2009). Despite increasing epidemiologic evidence of the risks of phthalate exposure to the developing male reproductive system, there are still debates regarding the risk of low-level exposures. For example, for almost two decades there have been debates about the mechanism of liver toxicity of one particular phthalate, diethyl hexyl phthalate (DEHP), which has resulted in some government bodies identifying the chemical as a possible human carcinogen, while others have determined that it is unclassifiable (often misinterpreted as meaning *not* carcinogenic).

In 1999, the European Commission temporarily banned the use of phthalates in children's toys, and in 2005 EU ministers voted unanimously to make the ban permanent (Lohse *et al.* 2003). These chemicals are now regulated under the EU's REACH directive (Registration, Evaluation, Authorization and Restriction of Chemicals). Three of the most common phthalates are included in the REACH list of Substances of Very High Concern. In contrast to the EU, rather than taking action, the US government in 1999 called for more research and held expert panels on six phthalates of concern (NTP 2000).

In 1998 the US Consumer Product Safety Commission (CPSC) released a report on risks to children from toys made of PVC containing the phthalate DINP (Babich 1998). The agency concluded that few children were at risk because the amount they may ingest would not be harmful. The study authors acknowledged that uncertainties existed and recommended that an exposure study be conducted to better determine the amounts of DINP that children may ingest during normal teething behavior. The CPSC did not recommend a ban on phthalates but encouraged the toy industry to voluntarily remove phthalates from rattles and teething toys pending additional scientific study. A further study by the CPSC concluded that oral exposure to DINP by mouthing plastic toys is 'not likely to present a health hazard to children' (Babich *et al.* 2004).

Because the US government was slow to act, state and local governments stepped in to regulate phthalates and some large retailers took action with their suppliers. In 2006, the city of San Francisco passed a law that banned phthalates in children's products. In 2007, the state of California passed a law to ban six phthalates in toys designed for children under age three. During the 2007–2008 legislative period, 29 states introduced bills to address phthlates and other chemicals of concern in children's products (Ekstrom 2008). Large retailers responded to this emerging patchwork of state and local legislation. In mid-2008 Walmart and Toys R Us put in place requirements for their suppliers to meet the EU phthalate restrictions (Kavilanz 2008). Finally, in August 2008 the *Consumer Product Safety Improvement Act* (CPSIA) was signed into law by President George W. Bush, permanently banning three phthalates in toys and children's products and provisionally banning three others, pending further study. For our purposes, a key element of this brief story is that phthalates were not regulated under existing statutes but rode the momentum of consumer outrage and media attention that followed a wave of product recalls in 2007, most notably Mattel's recall of millions of toys produced in China and painted with lead-based paint (Flaherty 2009; Becker *et al.* 2010). The federal regulation of phthalates did not emerge from a systematic regulatory regime or reflect the 'resolution' of scientific uncertainties relating to conventional notions of risk, but was made possible by a confluence of events that created a political opportunity to take action.

The contaminant 1, 4-dioxane is used in cosmetics and cleaning products. The US EPA has classified 1, 4-dioxane as a probable human carcinogen and the National Toxicology Program lists it as an animal carcinogen. This chemical is also a neurologic, renal, and respiratory toxin (Sullivan and Krieger 1999). Despite these determinations, the chemical is found in a wide range of consumer products, particularly cosmetics. A recent study by the Campaign for Safe Cosmetics found 1, 4-dioxane in bubble bath, bath wash, soap, lotion and shampoo (Sarantis *et al.* 2009). The chemical penetrates the skin. Because it is a contaminant produced during manufacturing, the FDA does not require listing it as an ingredient. Therefore, it is difficult to know what products contain it. Although the contaminant is found at low levels in some products, a large number of personal care products that are used daily may contain this substance, which in turn suggests that exposure may vary significantly among individuals.

In May 2009, a major Chinese supermarket chain pulled Johnson & Johnson baby products from its shelves because of safety concerns raised by the Campaign for Safe Cosmetics report (*China Daily News* 2009). Within days, the Shanghai Food and Drug Administration declared that their testing had found no contamination with 1, 4-dioxane (Associated Press 2009). Johnson & Johnson's response to this incident on their website was as follows: 'some of the ingredients in our products may contain 1, 4-dioxane as an incidental ingredient at extremely low levels. This trace ingredient is common in the personal care industry. . . . the low levels in some of our products present no risk to consumers.'

Manufacturers can add vacuum stripping to the manufacturing process to reduce 1, 4-dioxane to low levels. By designing the product with less harsh ingredients in the first place, 1, 4-dioxane can be eliminated altogether. Seventh Generation, a manufacturer of environmentally friendly cleaning and baby products requires their suppliers to vacuum strip to reduce levels of 1, 4-dioxane. The company's blog states that they are working to eliminate 1, 4-dioxane from their products. The EU and government of Canada have banned 1, 4-dioxane from cosmetics, but no action has been taken in the US.

## Science without argument

Recognizing the opportunity presented by corporate willingness to remove toxics from their products, corporations and environmental advocacy groups (sometimes working together) have constructed databases containing a wide range of toxicity data on chemicals (Schifano 2011). Large retailers like Walmart are requiring their suppliers to check these databases and either eliminate hazardous chemicals, or at a minimum inform the retailer when they are being used and why. These databases focus on *hazard* rather than *risk*. The distinction is important: hazard is an inherent characteristic of the chemical—its toxicity, deriving from its impact on biological systems. Risk is the probability that the chemical will actually cause disease to people, and involves considerations of the numbers of people who may be exposed, duration of exposure, and chemical concentrations. Determining risk also requires application of highly uncertain models specifying the 'dose-response curve'—how much chemical will cause how much risk. Federal rules like TSCA are risk-based in part because normal science is supposed to be able to determine if the benefits of regulation justify intervention in the economic market. Yet the increasing willingness of some corporations to take action—for example, by refusing to sell products with even modest evidence of hazard (such as evidence that a chemical causes cancer in mice)—shows that the market is far from homogenous when it comes to its view of risk. For large retailers like Walmart, there may be little economic downside to pressure suppliers to change the chemical make-up of products. This is another example of how the meaning of science is not enshrined in risk calculations, but is strongly determined by social context.

Science is not disappearing in this emerging new regime, but it is taking on a new role both in terms of its authority and how it is being brought into the process of negotiating decisions about chemical use. The new role of science in these non-regulatory approaches to reducing toxics is illustrated by the GreenScreen for Safer Chemicals, an assessment tool that helps 'industry, government and NGOs' to identify 'chemicals of high concern and safer alternatives'.[3] GreenScreen is a project of the not-for-profit organization Clean Production Action, which offers training in the GreenScreen method to educate designers and decision-makers in firms and organizations on how to implement the GreenScreen to compare and select safer chemicals for use in products and manufacturing processes. While GreenScreen is 'open, transparent, and publicly accessible', the method is sufficiently complex that GreenScreen also certifies 'GreenScreen Profilers' to provide technical assistance on a consulting basis to organizations that do not have the resources to carry out assessments themselves.

A second example is the Pharos project.[4] Pharos is an online database of toxicity information on building materials, accessible by subscription for $180 per year. It is designed to allow architects and builders to choose construction materials that minimize environmental and health hazards. The database (closely linked to the GreenScreen) summarizes data on health and environmental hazards for more than 22,000 chemicals and materials. The data come from more than 40 different governmental and professional society lists of different aspects of toxicity—for example, on animal carcinogens, mutagens and teratogens, on biopersistence, and on the tendency for a chemical to accumulate as it moves up food chains. These data are integrated into an overall score, through a fully transparent process. Pharos is a project of the Healthy Building Network, a not-for-profit organization funded by philanthropic organizations and subscriptions to the database. Pharos reports that more than 300 companies are using the database to make building design and construction decisions.

A third resource, aimed at reducing toxic chemicals in the workplace, is ChemHAT, a free online tool that draws from the Pharos database.[5] Organized and funded by several labor organizations and not-for-profit groups, including the BlueGreen Alliance and the United Steelworkers, it provides information about thousands of chemicals that are commonly found in the workplace. The ChemHAT website is extremely user friendly. One simply types in the name of a chemical on the home page, and a screen appears with simple, graphically clear information about health effects, regulatory status, environmental impacts, sources of exposure, and available alternatives. The information is color coded to indicate the strength of the effect and of the evidence, and selected background sources are provided as well.

## Implications for expertise, science and democracy, risk and chemicals policy

In the face of the failure of the US chemical regulatory regime to support systematic actions that limit human exposure to thousands of chemicals, a set of ad hoc, extra-governmental alternative arrangements seems to be emerging that is leading to restrictions in the production and use of selected chemicals that may pose risks to human and environmental health. At the same time, scientific data and assessments directly pertinent to decision-making about chemical use are also becoming available on web-based portals that allow individual as well as institutional users of chemicals to access information on human and environmental health effects.

These developments shine a clarifying light on the nature of risk, of science, and of the sources of cultural authority behind both. Decisions by corporations to take steps to limit the production and use of certain chemicals in the absence of government compulsion may in part be motivated by 1) a desire for predictability and uniformity in the face of multiple conflicting regulatory regimes; 2) a no-regrets approach where the costs of taking action are small given the potential legal or public relations consequences of not taking action; 3) a desire to help build corporate identity and business models around concern for consumer and environmental health; and 4) even a recognition that, especially for large retailers, there is the opportunity to exert significant leverage on how certain chemicals are used at little cost to them but with potentially significant benefits. Teasing out which of these motives (or others we have not identified) may be at play in which cases would require additional research. But whatever the motives, action is being taken without resolving the types of disputes that have gridlocked the formal regulatory regime at the federal level.

Social science research has long shown that assessments of risk and determinations of scientific fact are contextual and flexible in complex socio-technical systems (e.g. Douglas and Wildavsky 1982; Schwarz and Thompson 1990; Funtowicz and Ravetz 1990). Such insights have typically been applied to teasing out the power relations and negotiations around risk, science, and regulation, as well as the political essence of decision-making in that domain (Jasanoff 1990; Jasanoff and Wynne 1998). Here we want to suggest as well that the flexibility of risk and science are themselves actually providing opportunities for guiding alternative, ad hoc, informal governance arrangements that at least in part destabilize the standard political battle lines associated with the endless debates over formal regulatory regimes. The roles of scientists in these new arrangements are different as well. In particular, the expectation that scientists should speak with a unified voice via consensus panels or definitive studies that determine correct action is gone. When the science is post-normal, it adjusts to the available political arrangements. If an environmental non-profit organization and a corporation are battling over a regulatory standard, the science will never be good enough; if they are collaborating to find a path that suits them both, then the science is by definition good enough.

The value of post-normal science for guiding decisions about chemical use is further illustrated by the three decision tools we briefly described. Here, the users of the science are self-selecting. Companies and workers who are concerned about the adverse consequences of toxic chemicals in their products and their work environments can access and choose to act on information provided not by the standard sources of technical authority—the National Academies, government agencies, and so on—but by organizations that have assigned to themselves the task and responsibility of providing information for stakeholders that they can understand and use as a basis for action. The fact that this very same science might be endlessly contested in the courts if used to justify a formal regulatory action does not tell us if the information is more or is less reliable than the science purveyed by expert authorities; it tells us rather that such a judgment is immaterial and probably meaningless. The organization of the politics, and the reliability of the science, are mutually constitutive, and context-specific.

We are not arguing that the US does not need a functioning policy regime for chemical regulation. Nor do we want to suggest that the sort of alternative, informal, and self-organizing arrangements for taking action, and for supporting action with science that we have here described, are likely to be adequate to the challenge of governing the use of chemicals in society. Rather, we are impressed with the potential for creatively reorienting the politics of risk and the role of science in that politics. We have shown that there are opportunities for actors and institutions who are naturally divided by the standard regulatory regime to identify mutually acceptable actions. The negotiations involved in exploiting these opportunities do not depend on the expert and authoritative pronouncements of science. On the contrary, successful negotiation changes the nature of the science from the source and language of contestation to an additional justification for rational action. Publicly accessible databases and decision tools developed outside the standard institutions of scientific authority become legitimated sources of knowledge in support of the negotiated actions. Evidence that may be endlessly uncertain as normal science becomes 'good enough' as post-normal science, and authority that is supposed to be vested in a limited set of experts becomes broadly apportioned among stakeholders whose legitimacy derives not from specialized knowledge, but from effective participation in the negotiation process itself.

## Acknowledgements

We thank Jason Pearson for his extremely thoughtful and constructive comments on the first draft of this paper, and two anonymous reviewers for additional comments.

## Notes

1 Polychlorinated biphenyls, fully halogenated chlorofluoroalkanes, dioxin, asbestos, and hexavalent chromium. Cited in Appendix V of GAO's 'Chemical Regulation: Options Exist to Improve EPA's Ability to Assess Health Risks and Manage Its Chemical Review Program'. See GAO (2005).
2 *Corrosion Proof Fittings, et al. v. The Environmental Protection Agency and* William K. *Reilly, Administrator*, 947 F.2d 1201 (5th Cir. 1991).
3 Available at: www.cleanproduction.org/Greenscreen.php
4 Available at: www.pharosproject.net
5 Available at: www.chemhat.org/

## References

Associated Press. 2009. 'China Clears Johnson & Johnson Baby Products.' March 21.

Babich, M. 1998. 'The Risk of Chronic Toxicity Associated with Exposure to Diisononyl Phthalate (DINP) in Children's Products.' Bethesda, MD: US Consumer Product Safety Commission.

Babich, M.A., Chen, S-B., Greene, M.A. *et al.* 2004. 'Risk Assessment of Oral Exposure to Diisononyl Phthalate from Children's Products.' *Regulatory Toxicology and Pharmacology*, 40(2): 151–67.

Belliveau, M. 2011. 'The Drive for a Safer Chemicals Policy in the United States.' *New Solutions*, 21(3): 359–86.

Benachour, N. and Aris, A. 2009. 'Toxic Effects of Low Doses of Bisphenol-A on Human Placental Cells.' *Toxicology and Applied Pharmacology*, 241(3): 322–8.

Becker, M., Edwards, S. and Massey, R. 2010. 'Toxic Chemicals in Toys and Children's Products: Limitations of Current Responses and Recommendations for Government and Industry.' *Environmental Science and Technology*, 44(21): 7986–91.

Cannon, A. and Warner, J. 2011. 'The Science of Green Chemistry and Its Role in Chemicals Policy and Educational Reform.' *New Solutions*, 21(3): 499–517.

*China Daily News*. 2009. 'Supermarkets Suspends Sale of Johnson Products Amid Debate.' March 17.

CRS (Congressional Research Service). 2008. 'The Toxic Substances Control ACT (TSCA): Implementation and New Challenges.' Washington, DC: Congressional Research Service.

DiGangi, J., Schettler, T., Cobbing, M. and Rossi, M. 2002. 'Aggregate Exposures to Phthalates in Humans.' Washington, DC: Health Care Without Harm, July.

Douglas, M. and Wildavsky, A. 1982. *Risk and Culture*. Berkeley, CA: University of California Press.

Ekstrom, V. 2008. 'States Lead Feds in Toy Safety.' *Stateline*, April 2. Retrieved at: www.highbeam.com/doc/1G1-177442485.html

European Commission (EU). 2008. 'Updated European Risk Assessment Report on Bisphenol A.' Retrieved at: http://publications.jrc.ec.europa.eu/repository/bitstream/111111111/15063/1/lbna24588enn.pdf

Flaherty, E. 2009. 'Safety First: The Consumer Product Safety Improvement Act of 2008.' *Loyola Consumer Law Review*, 21(3): 372–91.

Funtowicz, S. and Ravetz, J. 1990. *Uncertainty and Quality in Science for Policy*. Dordrecht: Kluwer Academic.

Geiser, K. 2011. 'Redesigning Chemicals Policy: A Very Different Approach.' *New Solutions*, 21(3): 329–44.

Government Accountability Office (GAO). 2005. 'Chemical Regulation: Options Exist to Improve EPA's Ability to Assess Health Risks and Manage its Chemical Review Program.' GAO–05–458. Washington, DC, June. Retrieved at: www.gao.gov/assets/250/246667.pdf

Guston, D. 2006. 'On Consensus and Voting in Science.' In *The New Political Sociology of Science*. S. Frickel and K. Moore (eds). Madison, WI: University of Wisconsin Press, 378–404.

Hansson, S.O. 2002. 'Uncertainties in the Knowledge Society.' *International Social Science Journal*, 54(171): 39–46.

Heij, C. 1989. *Deterministic Identification of Dynamical Systems*. New York: Springer-Verlag.

Jacobs, M., Tickner, J. and Kriebel, D. 2010. 'Regulating Methylene Chloride: a Cautionary Tale about Setting Health Standards One Chemical at a Time' and 'Lessons Learned: Solutions for Workplace Safety and Health.' Lowell, MA: Lowell Center for Sustainable Production. Retrieved at: www.sustainableproduction.org/downloads/LessonsLearned-CaseStudy5.pdf

Jasanoff, S. 1990. *The Fifth Branch: Science Advisors as Policymakers*. Cambridge, MA: Harvard University Press.

Jasanoff, S. and Wynne, B. 1998. 'Science and Decision Making.' In *Human Choice and Climate Change, vol. 1: The Societal Framework*. S. Rayner and E. Malone (eds). Columbus, OH: Battelle Press, 1–87.

Kavilanz, P. 2008. 'Wal-Mart, Toys 'R' Us Unveil New Safety Rules.' CNNMoney, February 15. Retrieved at: http://money.cnn.com/2008/02/15/news/companies/toysafety_update/

Kriebel, D. 2007. 'The Reactionary Principle.' *Occupational & Environmental Medicine*, 64: 573–4.

Kriebel, D. 2009. 'How Much Evidence is Enough? Conventions of Causal Inference.' *Law & Contemporary Problems*, 72: 121–36.

Kriebel, D., Tickner, J., Epstein, P. *et al.* 2001. 'The Precautionary Principle in Environmental Science.' *Environmental Health Perspectives*, 109(9): 871–6.

Kuhn, T.S. 1962. *The Structure of Scientific Revolutions*, 1st edn. Chicago: University of Chicago Press.

Lang, I., Galloway, T., Scarlett, A. *et al.* 2008. 'Association of Urinary Bisphenol A Concentration with Medical Disorders and Laboratory Abnormalities in Adults.' *Journal of the American Medical Association*, 300(11): 1303–10.

Layton, L. and Lee, C. 2008. 'Canada Bans BPA from Baby Bottles.' *The Washington Post* April 19. Retrieved at: www.washingtonpost.com/wp-dyn/content/article/2008/04/18/AR2008041803036.html

Lohse, J., Wirts, M. and Ahrens, A. *et al.* 2003. 'Substitution of Hazardous Chemicals in Products and Processes.' Compiled by Ökopol GmbH and Kooperationsstelle Hamburg for the European Union Directorate General Environment, Nuclear Safety and Civil Protection of the Commission of the European Communities. Hamburg, Germany, March.

McGinn, A. 2000. 'Why Poison Ourselves? A Precautionary Approach to Synthetic Chemicals.' Paper 153. Washington, DC: Worldwatch Institute, November.

National Conference of State Legislatures (NCSL). 2014. 'NCSL Policy Update: State Restrictions on Bisphenol A (BPA) in Consumer Products.' Retrieved at: www.ncsl.org/research/environment-and-natural-resources/policy-update-on-state-restrictions-on-bisphenol-a.aspx

National Toxicology Program (NTP). 2000. 'NTP-CERHR Expert Panel Reports on Di (2 ethylhexy) Phthalate, Butyl Benzyl Phthalate, Di-n-butyl Phthalate, Di-isononyl Phthalate, Di-n-octyl Phthalate, and Di-isodecyl Phthalate.' Research Triangle Park, NC: US Department of Health and Human Services.

National Toxicology Program (NTP). 2008. 'Draft NTP Brief on Bisphenol A.' Research Triangle Park, NC: US Department of Health and Human Services. Retrieved at: https://ntp.niehs.nih.gov/ntp/ohat/bisphenol/bisphenol.pdf.

National Toxicology Program (NTP). 2011. 'Report on Carcinogens', 12th edn. Research Triangle Park, NC: US Department of Health and Human Services, Public Health Service, National Toxicology Program, 499. Retrieved at: http://ntp.niehs.nih.gov/?objectid=03C9AF75-E1BF-FF40-DBA9EC0928DF8B15

Occupational Safety and Health Administration (OSHA). 1997. 'Final Standard for Methylene Chloride.' Washington, DC: US Department of Labor. Retrieved at: www.osha.gov/SLTC/methylenechloride/

Peretz, J., Vrooman, L., Ricke, W.A. *et al.* 2014. 'Bisphenol A and Reproductive Health: Update of Experimental and Human Evidence 2007–2013.' *Environmental Health Perspectives*, 122(8): 775–86.

Pielke Jr, R. 2007. *The Honest Broker: Making Sense of Science in Policy and Politics*. Cambridge: Cambridge University Press.

Rayner, S. and Cantor, R. 1987. 'How Fair Is Safe Enough? The Cultural Approach to Societal Technology Choice.' *Risk Analysis*, 7(1): 3–9.

Rossi, M., Thorpe, B. and Peele, C. 2011. 'Businesses and Advocacy Groups Create a Road Map for Safer Chemicals: The Bizngo Principles for Chemicals Policy.' *New Solutions*, 21(3): 387–402.

Rust, S., Fauber, J. and Kissinger, M. 2009. 'Baby Bottle Makers Agree to Stop Using Bisphenol A.' *Journal Sentinel*, Milwaukee, March 6.

Sarantis, H., Malkan, S. and Archer, L. 2009. 'No More Toxic Tub: Getting Contaminants out of Children's Bath and Personal Care Products.' Campaign for Safe Cosmetics, March.

Sarewitz, D. 2004. 'How Science Makes Environmental Controversies Worse.' *Environmental Science and Policy*, 7(5): 385–403.

Schierow, L-J. 2013. 'The Toxic Substances Control Act (TSCA): A Summary of the Act and its Major Requirements.' Washington, DC: Congressional Research Service. Retrieved at: www.fas.org/sgp/crs/misc/RL31905.pdf

Schifano, J. 2011. 'A Vision for Safer Chemicals: Policy, Markets, Coalitions, and Science. Introduction to a Special Issue on Designing a Chemically Safer Future.' *New Solutions*, 21(3): 323–7.

Schwarz, M.L. and Thompson, M. 1990. *Divided We Stand: Redefining Politics, Technology and Social Choice*. Philadelphia, PA: University of Pennsylvania Press.

Sissell, K. 2009. 'Sunoco Voluntarily Restricts BPA; Lawmakers Propose a Ban.' *IHS Chemical Week*, March 17. Retrieved at: www.chemweek.com/markets/engineering_plastics/polycarbonate/Sunoco-Voluntarily-Restricts-BPA-Lawmakers-Propose-a-Ban_17644.html

Stirling, A. 2007. 'Risk, Precaution and Science: Towards a More Constructive Policy Debate.' *European Molecular Biology Organization Reports*, 8(4): 309–15.

Sullivan, J. and Krieger, G. (eds). 1999. *Clinical Environmental Health and Toxic Exposures*, 2nd edn. Philadelphia, PA: Lippincott Williams & Wilkins.

Swan, S., Liu, F., Hines, M. *et al.* 2009. 'Prenatal Phthalate Exposure and Reduced Masculine Play in Boys.' *International Journal of Andrology*, 32(2): 259–69.

Thomas, K. 2014. 'The "No More Tears" Shampoo, Now with No Formaldehyde.' *New York Times*, January 18, A1. Retrieved at: www.nytimes.com/2014/01/18/business/johnson-johnson-takes-first-step-in-removal-of-questionable-chemicals-from-products.html?_r=0

US Centers for Disease Control and Prevention. 2009. 'The Fourth National Report on Human Exposure to Environmental Chemicals.' Washington, DC: US Department of Health and Human Services. Retrieved at: www.cdc.gov/exposurereport/

US Food and Drug Administration. 2014. 'Questions & Answers on Bisphenol A (BPA) Use in Food Contact Applications.' Retrieved at: www.fda.gov/Food/IngredientsPackagingLabeling/FoodAdditivesIngredients/ucm355155.htm

US Environmental Protection Agency (USEPA). 2012a. 'Existing Chemicals Program: Strategy.' February. Retrieved at: www.epa.gov/oppt/existingchemicals/pubs/Existing_Chemicals_Strategy_Web.2–23–12.pdf

US Environmental Protection Agency (USEPA). 2012b. 'Essential Principles for Reform of Chemicals Management Legislation.' Retrieved at: www.epa.gov/opptintr/existingchemicals/pubs/principles.html

Vogel, S. 2009. 'The Politics of Plastics: The Making and Unmaking of Bisphenol A Safety.' *American Journal of Public Health*, 99(S3): 559–66.

vom Saal, F., Timms, B., Montano, M. *et al.* 1997. 'Prostate Enlargement in Mice Due to Fetal Exposure to Low Doses of Estradiol or Diethylstilbestrol and Opposite Effects at High Doses.' *Proceedings of the National Academy of Science*, 94(5): 2056–61.

# 8 In search of certainty

## How political authority and scientific authority interact in Japan's nuclear restart process

*Paul J. Scalise*

### Introduction

Japan's regulators are striving to come to grips with a new and increasingly skeptical political landscape as they slowly restart the fleet of nuclear reactors in the wake of the Fukushima Dai-ichi Nuclear Power Plant accident. In addressing these challenges, the concept of the 'independent regulator' figures prominently in the *Act for Establishment of the Nuclear Regulation Authority* since 2012.[1] Promising to balance the demands of nuclear safety with national security and economic growth,[2] decision-makers have organized ministerial advisory councils and committees to deal with further complexities of policy and safety. Other industrial democracies around the world have also been struggling with perceived inadequacies in nuclear and electric power market regulation since the accident at Three Mile Island in 1979 and Chernobyl in 1986. In the United States, for example, the Nuclear Regulatory Commission—though an independent agency since 1974—has nevertheless continued to encounter a series of allegations about its perceived bias towards the nuclear industry (Byrne and Hoffman 1996; Von Hippel 2011).

Heretofore, scholarship on the theories of regulation has been broadly categorized into two main approaches: public interest models and regulatory capture models. The first model of the regulatory process, the public interest approach, typically asserted that markets were (and are) generally too fragile and inefficient to be left alone. With government regulation virtually costless, regulators—albeit imperfect and fallible human beings—intervened in attempts to correct inefficiencies for the sake of public interest (Priest 1993). Researchers, offering a variation of this public interest model in the East Asian context, proffered the 'developmental state' or 'plan rational' model of regulatory oversight (Bagchi 2004; Greene 2008; Johnson 1982; Woo-Cumings 1999). Authors argued that the Japanese regulator, in the words of one of its most well-known proponents, 'alter[ed] market incentives, reduc[ed] risks, offer[ed] entrepreneurial visions, and manag[ed] conflicts' (Johnson 1999: 48). This variation on the public interest model was common until the mid-1990s, when a series of scandals and rival theories emerged in the wake of the so-called 'lost decades' of Japan's economic and political development to challenge assumptions of decision-making power in the government-business relationship.

The second model of regulatory control posited a relationship similar to a material transaction. Arguing against the public interest model, theorists like George Stigler (1971) moved beyond assumptions of selfless decision-makers engaged in apparent acts of public welfare towards a more 'economic' explanation of regulation. This model resurrected, albeit in modified form, the nineteenth-century Marxist contention that political institutions were actually bulwarks for 'monopoly capital' seeking to justify, reinforce, and protect the government-business status quo (Marx 1970). In this newly reworked theory in which the regulated 'capture' the regulator, protecting the material self-interests of large corporations was not necessarily the primary objective. Rather, the theory suggested how a much broader dynamic involving various organized economic interests—large and small—came together in the political marketplace of supply and demand. The 'supply' represented politicians and bureaucrats wishing to 'sell' their services to the highest bidder in the form of favorable laws and regulations. The 'demand' represented their respective primary constituencies looking to 'buy' these services in the form of superior organization, information collecting, campaign cash contributions, and electoral votes (Friedman 1993; Peltzman 1993; Stigler 1971; Stigler and Friedland 1962). Scholars such as Mark Ramseyer and Francis Rosenbluth (1993), Richard Katz (1998) and Aurelia George Mulgan (2002) among many others have—consciously or not—followed in Stigler's theoretical footsteps by applying this regulatory capture model to Japan's political economy. Indeed, Kiyoshi Kurokawa, Chairman of the National Diet's official report on the nuclear accident at Fukushima, argued that the accident could have (and would have) been avoided were it not for the 'regulatory capture' of Japan's nuclear industry creating what was effectively a 'man-made disaster' (Kurokawa *et al.* 2012).

Notwithstanding Kurokawa's contention, scholarly work on the theories of regulation and decision-making processes has attempted to stretch beyond the static dichotomy of the public interest/regulatory capture models. As legal scholar Richard A. Posner (1974: 356) pointed out, 'Neither theory can be said to have . . . substantial empirical support. Indeed, neither theory has been refined to the point where it can generate hypotheses sufficiently precise to be verified empirically'. Decision theorists in political science, long skeptical of the underlying assumptions characterizing fully rational decision-makers in both approaches to regulatory theory (i.e. goal-oriented actors who have fixed, transitive preferences for alternatives in which cost-benefit analyses maximize either one's material self-interest or diffuse consumer welfare gains) began to conceptualize alternative models rooted in psychology, ideas, and behavioral choice to explain puzzling policy outputs (Campbell 1992; Derthick and Quirk 1985; Quirk 1988).

This chapter takes as its focus three specific cases of the government–business relationship as it applies to the debate over regulatory authority and its underlying purpose. These cases involve Japanese electric power companies whose nuclear reactors were shut down in local host communities for regularly scheduled maintenance checks following the nuclear crisis at Fukushima Dai-ichi Nuclear power plant on March 11, 2011. Yet, for us the important question is the extent to which *variation* in scientific consensus regarding nuclear safety, or the manner in which scientific authority is institutionalized in Japanese society to answer such questions, affect policy outcomes and the intervention of political authority under socioeconomic circumstances that are otherwise similar. That is to say, do Japanese actors behave differently than they would if they were just pursuing their material self-interest in a narrowly utilitarian sense?

I begin this chapter by briefly discussing the evident weaknesses of the two conventional models—the public interest model and the regulatory capture model—as they are applied to Japan's electric power/nuclear industry. This point of comparison—that regulators are either selfless technocrats on one extreme or self-interested material maximizers on the other extreme—allows us to formulate the two null hypotheses that are central to the research design of this chapter. I then propose to advance a different conceptual framework for understanding the puzzling socioeconomic situation of the Japanese energy market based on what Judith Goldstein and Keohane (1993) call 'causal beliefs'—beliefs about cause–effect relationships that derive their authority from the shared consensus of recognized scientific elites. Then, I will investigate the variation in three similar nuclear restart stories in Japan with strikingly different outcomes: Tomari No. 3, Ōi Nos. 3–4, and Tsuruga No. 2. These three case studies of varying nuclear restarts (the dependent variable) illustrate how causal beliefs provide guidance to actors to achieve their objectives in frenetic moments of uncertainty. I will conclude with some observations on the Japanese government–business relationship and the power of scientific authority in Japan.

## The weaknesses of conventional models

At first glance, the conventional explanation of government oversight of the nuclear power industry is consistent with the public interest model of regulation, which 'holds that regulation is supplied in response to the demand of the public for the correction of inefficient or inequitable market practices' (Posner 1974: 335). Since the enactment of the *Atomic Energy Basic Law* in 1955 limiting the uses of nuclear power, Japan has placed a growing number of regulatory requirements on the owners and operators of nuclear power plants. The law establishing the National Diet of Japan Fukushima Nuclear Accident Independent Investigation Commission (2011), for example, was tasked with investigating the background and causes of the Fukushima nuclear accident, so it may never happen again; the *Act for the Establishment of the Nuclear Regulation Authority* (2012) created a new and independent regulatory authority replacing its discredited predecessor, the Nuclear Safety Commission (NSC), with an agency located at the Ministry of Environment (MOE); and the *Act on the Regulation of Nuclear Source Material, Nuclear Fuel Material and Reactors* was revised multiple times to ensure stricter safety laws and penalties for operational compliance failure.[3] These examples and 206 additional laws, political ordinances, and basic regulations like them were enacted following the Fukushima nuclear crisis in 2011—the highest number passed in Japan's post-war history (Figure 8.1).

Upon closer examination of the evidence, however, one notices a lack of high regulatory activity in Japan from 1955 to 1985, despite nuclear incidents at home and abroad.[4] McLaughlin *et al.* (2000) document 41 criticality accidents at process facilities carrying out operations with fissile material between 1955 and 1985. In Japan, pre-Fukushima nuclear power criticality incidents and accidents occurred six times between 1955 and 1985. Were the public interest model of regulation correct, one should have seen an increase in regulatory enactment and revision following these incidents. We do not. Indeed, in 1975, 1978, 1979, and 1981 (twice), one notices little activity at all despite domestic nuclear criticality incidents.

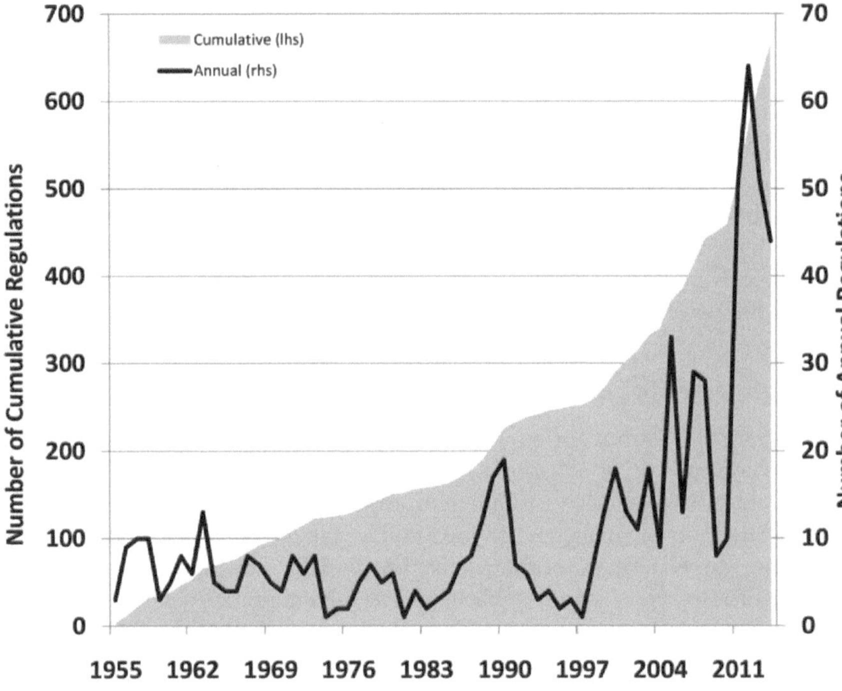


*Figure 8.1* Nuclear power regulation enactments (1955–2014)

Notes: 'Regulations' include the total number of nuclear-related laws, treaties, political ordinances, notifications, regulations, and edicts enacted in Japan since 1955. 'Cumulative' refers to the total existing number on the books in a given year less the same number of regulations repealed. 'Lhs' = left-hand side. 'Rhs' = right-hand side.

Source: *Nihon horei sakuin* [Japan Law and Regulation Index]. Retrieved at http://hōurei.ndl.go.jp/SearchSys/frame/seitei_top.jsp.

There has been considerable journalistic speculation on how a potential conflict of interest within the regulated industry and its regulator, the Ministry of Economy, Trade, and Industry (METI), might have resulted in a lapse in effective oversight, thus producing the Fukushima nuclear accident.[5] To be sure, the public continues to harbor a lack of confidence in the government-business relationship, especially with respect to the energy sector (Hommerich 2012). These inconsistent regulatory developments in reaction to critical incidents and accidents, in conjunction with the public's lack of confidence in the government–business relationship, suggests that the empirical support of the public interest model is relatively weak or at least inconsistent as defined by Posner.

The more commonly applied model to explain Japan's government–business relationship in the nuclear industry, however, is one of regulatory 'capture'—an explanation with militaristic overtones. The regulated effectively conquer the regulator, but how they accomplish this task varies by author. Jeff Kingston (2012) argues that Japan's energy policy is dominated by a 'nuclear village' in which pro-nuclear institutional and individual forces span like a spider web across the public and private sectors: 'the utilities, nuclear vendors, bureaucracy, Diet (Japan's parliament), financial sector, media and academia.' In Kingston's view, this 'nuclear village' is a powerful, unified interest group with a specific agenda: profit. Kenji Kushida (2012) repeats the 'nuclear village' argument but places greater emphasis on Japan's power companies. '[They] are regulated in a way that guarantees profit . . . As a result, the industry enjoys an extreme concentration of financial resources that fund not only suppliers, but a wide range of activities.' Charles Ferguson and Mark Jansson (2013) contend that nuclear regulations are

> tailored to serve companies' business interests in Japan and by weak enforcement of whatever rules are on the books . . . Instead of aggressively pursuing measures to reduce risks, the nuclear village aggressively promoted a 'safety myth' that the country's nuclear power reactors were already one-hundred per cent safe.

All adherents to the regulatory capture model, despite their variations in emphasis, implicitly agree that the private sector comprises rational, self-interested individuals or groups seeking regulation deliberately as a means of maintaining (or enhancing) profitability. When regulation has the potential for directly (and negatively) affecting the nuclear industry, the argument is—actually, it is more of an assumption—that the regulation was designed to serve the industry's interest rather than the public interest.

These arguments, however, also lack a theoretical foundation and exhibit explanatory weaknesses. First, they do not tell us under what conditions interest groups succeed or fail in obtaining favorable regulations or judgments, how representative these reported incidents alleging conflict of interest in the nuclear industry are, or where the explanatory power may be if the 'captured' government agency regulates separate industries having conflicting interests.

Second, consumers of electric power (including nuclear power) have an obvious interest in the outcome of the regulatory process—why might they not be able to

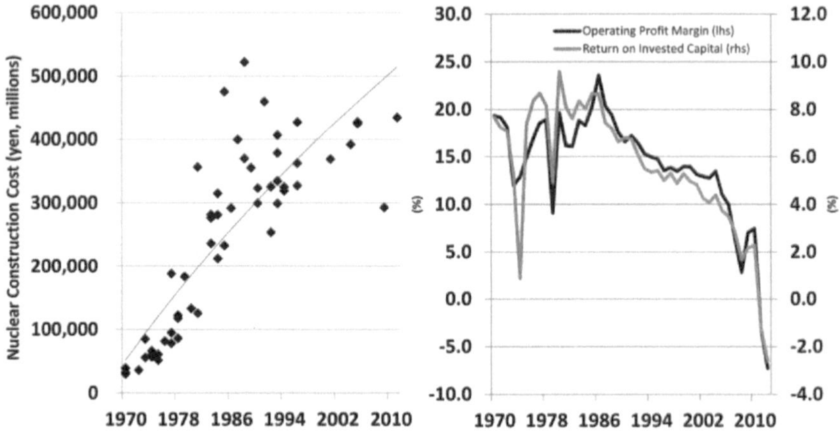


*Figure 8.2* Declining profits, declining returns, and rising nuclear construction costs (1970–2012)

Notes: Operating profit, also known as earnings before interest and tax, is the profit that is earned from a company's core business activities such as electric power generation. This value does not include any profit from its outside investments. The Operating Profit Margin, or OPM, is calculated as (Operating Profit/Revenues) * 100; Return on Invested Capital is calculated as (Net Operating Profit/Invested Capital) * 100. 'Lhs' = left-hand side. 'Rhs' = right-hand side.

Sources: *Denryoku shinsetsubi yōran [Handbook of Electric Power New Facilities]* (various years); Federation of Electric Power Companies financial database; chart created by Paul J. Scalise.

'capture' the regulator and government using political resources, too? Rigorous econometric studies in other countries and sectors have found little evidence that political resource endowments such as cash contributions to political parties, advertising budgets, and other tools exert much (if any) impact on policy outputs favoring one interest group or another (Baumgartner *et al.* 2009; Chappell 1982; Grenzke 1989; Mayer 1991; Sorauf 1992; Welch 1982). Testing for the impact of political resources on liberalization of Japan's electric power market, Scalise (2013) also finds little empirical support of political resources 'buying' favorable government policies. If cash contributions, advertising budgets, and close political connections resulted in favorable regulations of incumbent power suppliers, the voting record in standing committees should have reflected that. It does not.

Third, even if compelling support existed for the notion that political resources effectively 'buy' public policy (and, as pointed out, the evidence is weak), the fundamental proposition that rent-seeking behavior in Japan's energy sector leads to a regulatory environment favoring profit maximization of the rent-seeker is weaker still. Figure 8.2 plots rising construction costs, falling profitability, and falling returns of the owners and operators of nuclear power plants since 1970. Japan's construction costs for nuclear power plants rose from roughly ¥29.8 billion in 1970 to almost ¥435 billion in 2011. Analysts find a direct positive correlation between increased safety regulations and increased Japanese construction costs over time (Dauvergne 1993; Marshall and Navarro 1991). As construction costs of essential facilities rise, power companies face a regulated tariff system that also constrains the ability to pass along rising expenditures in a timely and efficient manner. Consequently, the average profit margin of the electric power industry—the profit ratio used to measure cost control—has continued to fall from 24 percent in 1986 to –7 percent in 2012, while the return on invested capital—the performance ratio used to measure how much return is generated from every Japanese yen—has continued to fall from 9 percent in 1982 to –2.6 percent in 2012 (Figure 8.2). In short, the empirical evidence shows that the owners and operators of essential power-generating facilities in Japan have lost control of their operating expenditures and profitability over extended periods of time.

Scholars often point to the difficult and costly ways in which the power companies and the central government must navigate politically hostile local communities, skeptical media-driven public perceptions, and the complexities of law in order to successfully site, license, and construct essential facilities in Japan (Aldrich 2008; Broadbent 1998; Lesbirel 1998; Sakai 2012). This valuable empirical research, however, while helping indirectly to explain rising prices and declining profits of incumbent power suppliers, does not support the contention of regulatory capture model(s) that institutions are stacked in their corporate favor. Indeed, why would supposedly powerful interest groups who are thought to manipulate the political process for their material self-interest tolerate—let alone work within—institutions that encourage genuine public interest considerations in the decision-making process? Eminent domain, the right to protest, private property rights, universal service,[6] and other laws and legal rights pertaining to Japan's energy sector ironically contribute to decreased shareholder value, consumer welfare losses, and declining corporate profitability over time.

## Scientific consensus, causal beliefs, and the politics of expertise

If Japanese regulators of the electric power/nuclear industry are not always strategically competent technocrats working in the best interest of the public[7] nor 'captured' regulators of the regulated industry itself,[8] one still needs an alternative causal model to explain not only the puzzling behavior surrounding the Fukushima nuclear accident (i.e. why did regulators fail to address the back-up diesel generator flooding and potential nuclear meltdown risk in advance?), but also the counter-intuitive characteristics of the Japanese electric power industry in failing to defend its own material self-interest (i.e. why were the regional monopolies and nuclear power owners and operators *still* plagued by rising construction costs, constrained prices, and falling profits for decades?). To be sure, these two seemingly distinct puzzles are actually intertwined; they are both arguably rooted in flawed assumptions of rational choice theories of the decision-making literature in explaining regulatory outputs and market behavior in Japan. Their explanatory weaknesses lead us to reconsider the role of policy subsystems and the processing of issues.

As decision theorist James March (1994: 9) wrote: 'Although decision-makers try to be rational, they are constrained by limited cognitive capacities and incomplete information, and thus their actions may be less than completely rational in spite of their best intentions and efforts.' This bounded rationality, or what political scientist Bryan D. Jones (2001) calls an actor's 'cognitive architecture', often results in actors systematically failing to meet expected utility calculations in a rational decision-making model of actor preferences (Camerer 1998; Camerer and Thales 1995; Thaler 1991, 1992). Short-term memory traits, attention shifts, and constrained search strategies often produce inadequate tradeoffs analyses before decisions are reached, thus generally reinforcing the policy subsystem's status quo. The result is actors relying ever more on disinterested third-party authorities and institutions for guidance (Jones 1994).

This situation is abundantly evident in the case of Japan. For most of the post-war era, political authority—constrained by limited time, energy, and resources in addressing technical issues—considered scientific authority a pivotal policy making component in answering the question of resource poor Japan's energy challenges. Policy subsystems and their regulatory apparatus relied heavily on third party experts as a buttress and a legitimizer of drafted and enacted policy. Unlike the US or the UK, Japan's legislators were (and are) not equipped with large staff and expertise (Kodderitzsche 1991). Each legislator has at most three staffers to aid in administrative duties and other policy related matters, putting the weight of legislative vetting and drafting on the respective ministry from which the bill usually originates (Scalise 2009). Consequently, the post-war rise of the standing advisory council, or *shingikai*, in conjunction with other independent regulatory bodies, provides formal advice and oversight to the Ministry and Japanese Diet. In addition to being a check on agency and political action, these *shingikai* and third-party expert panels legitimize decisions by giving them the imprimatur of a group of respected authorities outside the bureaucracy and political realm (Schwartz 2001).

In the next section, I offer a demand-side analysis of Japanese decision-making that turns away from prior models' narrow focus on how assumed fixed, transitive preferences of incumbent actors with political resources vie to influence the political market for public interest considerations (one extreme) or their material self-interest (another extreme). Rather, it follows in the footsteps of Judith Goldstein and Robert Keohane (1993), Bryan Jones (2001), and Cornelia Woll (2008). It argues that the decision-making process, in the midst of exogenous shocks that created moments of socio-political uncertainty, becomes heavily reliant on the politics of expertise to regain institutional stability and certainty. When scientific authority is unanimous in its judgment, political authority rarely intervenes in the name of economic necessity. When scientific authority is divided in its judgment, political authority often seizes upon what John Kingdon (1984) calls 'windows of opportunity' in the decision-making process.

## Tomari No. 3 (scientific authority agrees: safe)

The first case in Japan's nuclear restart analysis is the puzzling example of Tomari No. 3, owned and operated by Hokkaido Electric Power Co (HEPCO). After a five-month battle to be declared operational by their regulator, the (then) Nuclear and Industrial Safety Agency (NISA), Tomari No. 3 was the first nuclear power plant technically to restart its post-Fukushima commercial operations on August 17, 2011. The victory—albeit short lived—underscored the ability of scientific authority in the form of 'stress tests'[9] to provide temporary cover against the growing economic pressures from the political sphere to restart the reactors in the midst of general safety concerns.

Located in the northwestern village of Tomari in Hokkaido Prefecture, Tomari No. 3 was the third and final nuclear reactor to be built not only in Hokkaido but also the last nuclear reactor to be brought commercially online in Japan before the tragic events of 3.11. The model was a pressured water reactor, or PWR, with an installed capacity of 912 megawatts (MW), costing ¥470,000 per kilowatt (US$4835, US$1=¥97.2) to build, and taking nine years to complete from its initial site approval in October 2000 until the start of its commercial operation in December 2009.

Since the reorganization of the electric power sector in 1951, HEPCO was (and is) the dominant supplier of the Hokkaido region. Like all general electric utilities, or *ippan denki jigyōsha*, HEPCO remained a vertically integrated and privately owned electric power company that combines its generation, transmission, and distribution assets within one corporate entity. Acting as a 'natural monopolist', it generated in FY2010 72 percent of all kilowatt-hours (kWh) sold in its service region with new entrants (2.9 percent), the wholesale supplier J-Power (1.6 percent), off-grid municipal generators (3.5 percent), and off-grid independent power producers (19.8 percent) generating the balance of supply.

Nuclear power came to hold a special place in the Hokkaido region, accounting for approximately 54 percent of HEPCO's total generation output on a kWh basis. From an economic perspective, Tomari NPS was also the lifeblood of the municipality. Tomari—an isolated and sparsely populated village at the base of the Shakotan Peninsula with a population of only 1,100—relied on Tomari NPS for 30 percent of the town's employment, 80 percent of the town's tax revenues, and an additional ¥1.8 billion (US$18.5 million, US$1=¥97.2) in electric power development subsidies for the region coming from the central government.

After going offline for regularly scheduled maintenance from January 5, 2011, Tomari No. 3 entered its test adjustment phase on March 7, 2011—four days before the 'Great East Japan Earthquake'. A typical 'adjustment phase' usually lasts no more than one month in Japan as engineers analyze normal operations for inconsistencies. However, in March–August 2011, Tomari No. 3 was effectively in a state of regulatory limbo, neither receiving government approval to enter its final commercial phase of operations nor being ordered to shut down completely. Consequently, the uncertainty surrounding the nuclear reactor left open the possibility that the utility would be forced to halt operations again, throwing in doubt the schedule for the next regularly scheduled maintenance shutdown required every 13 months from the start of commercial operations.

During this extended period, new procedures were introduced in Europe regarding stress tests on nuclear power plants to assess how far reactors could withstand major earthquakes, tsunamis, and other natural disasters. Although the International Atomic Energy Agency (IAEA) Director General Yukiya Amano asked all IAEA member states to consent to an agency-led peer review of these new reactor safety tests and nuclear regulatory reviews in the wake of the Fukushima nuclear crisis (Maclachlan 2011), it was not until Prime Minister Naoto Kan instructed NISA to order similar 'stress tests' on Japan's nuclear reactors that the five-month stalemate reached its boiling point.

Before June 2011, the presumption of nuclear restarts was evident. HEPCO had long assumed nuclear restarts in its annual business projections. President Yoshitaka Sato reassured the public that HEPCO would invest approximately ¥300 billion (US$3bn, 1US$=¥97.2) on safety measures at Tomari NPS over the coming years in the hope that safety concerns would ease. Hokkaido's gubernatorial election in April 2011—a usual testing ground for gauging public sentiment—managed to attract three newcomers to Hokkaido politics (i.e. Kimura Toshiaki, Satoshi Miyauchi, and Katsuya Tadashi) with anti-nuclear aspirations. Yet despite their best efforts, anti-nuclear issues barely provoked a serious challenge to Hokkaido's political status quo. Two-time incumbent Governor Harumi Takahashi, a former METI bureaucrat, refused to address nuclear power as a campaign issue and won in a landslide victory with 69.4 percent of the vote (*Asahi Shimbun*, April 12, 2011: 1). 'Anxieties run high', Shizuo Mishima Secretary General of the Tomari Chamber of Commerce confessed, 'but the nuclear power plant is already up and running, and without it [Tomari] village cannot survive' (*Asahi Shimbun*, April 20, 2011: 30).[10]

After June 2011, these public 'anxieties' became increasingly prominent, as the nuclear crisis at Fukushima Dai-ichi NPS was ongoing with growing concerns that the earthquake itself might have had a more direct role in the meltdown; street protests ensued attracting a small, but noticeable following. On June 19, 2011, Sapporo City lawyers and university professors opened exploratory meetings for the possibility of suing the central government lest it grant Tomari No. 3 its commercial status (*Asahi Shimbun*, June 20, 2011: 23).

The key delay in moving from the 'adjustment phase' to the 'commercial phase' in Tomari No. 3's status was the widespread fear that nuclear power plants were inherently unsafe. Nationwide public opinion polls shifted dramatically between March and June 2011 with 74 percent in favor of eventual phase-out (*Asahi Shimbun*, June 14, 2011: 1). Prime Minister Kan, too, having experienced a road to Damascus conversion in Japan's energy policy, viewed nuclear power as too dangerous to continue unchecked and advocated its gradual phase-out. In a press conference held on July 12, 2011, Kan arrived at a compromise to assuage the national mood and deal with the idled reactors. He announced a two-phase nuclear safety examination. The first phase would have the power companies carry out stress tests using computer models to determine how their plants would perform in the event of natural disaster. Having passed those tests, winning over the local authorities for restart approval would be the responsibility of the companies. The second phase would import standards used by the European Union to conduct post-Fukushima stress tests and incorporate the additional findings from Fukushima to the upgrading of all reactors in Japan. Because of its unusual status, however, Tomari No. 3 would be exempt from phase two tests. NISA had long considered the reactor's test status as problematic and instructed HEPCO to undergo the final phase-one inspections as quickly as possible to win over the central government—a move that HEPCO was happy to oblige given the rising fossil fuel costs on the world spot market.

Since gaining increased decision-making independence in the 1990s, the prefectural governors played a significant role in local politics (Horiuchi 2009). The siting, licensing, and construction of power plants were no exception to that increased power. Unlike future reactor siting, however, the Hokkaido prefectural government began pressuring the Kan Cabinet for nuclear restarts in Hokkaido starting on July 15, 2011. METI Minister Banri Kaieda put the ball back into the Hokkaido government's court by insisting that that the prefectural government sell the reliability of the stress tests to the local municipalities. With polls indicating 51 percent willing to accept the safety measures required by the central government in order to reactivate nuclear reactors (35 percent against), Gov. Takahashi announced the public was satisfied with the stress tests. Tomari No. 3 was reactivated on August 17, 2011. To be sure, public protests continued in the coming months with limited reactions in Hokkaido. The class-action lawsuit expecting to draw at least 1,000 plaintiffs eventually amounted to only 38 participants (*Asahi Shimbun*, August 2, 2011: 33). Public protests in the streets expecting thousands, ultimately reported only dozens. And for a while, it seemed that Minister Kaieda's dictum regarding the nature of nuclear restarts proved prescient: 'This is all technical procedure', he insisted, 'not a decision to be made politically'.

## Ōi Nos. 3–4 (scientific authority divided: safe or unsafe?)

The second case of Japan's nuclear restarts involves the puzzling debates surrounding Ōi Nos. 3–4, owned and operated by Kansai Electric Power Corporation (KEPCO). Were the Ōi Nos. 3–4 reactors sitting upon or close to active fault lines or not? If the case of Tomari No. 3 emphasizes the power of scientific consensus under METI's NISA to provide political cover and a degree of certainty to the decision-making process, the lack of scientific consensus among disinterested experts regarding the possibility of active fault lines in the Kansai region highlights how quickly political in-fighting can blossom, leading to debates about economic necessity in the form of restarts.

Like HEPCO, KEPCO faced a similar declining socioeconomic situation in the Kansai region partially as a result of their pre-Fukushima reliance on nuclear (Table 8.1). Both nuclear operators had similar political resource endowments, had faced not-in-my-backyard (NIMBY) siting resistance for decades, and were experiencing declining profitability over time. Restarts were essential for both companies' financial profiles. In addition, like Tomari No. 3, Ōi Nos. 3–4 were both expensive pressurized water reactors (PWRs) built in an equally small town of Ōi (Population: 9,216), Fukui Prefecture, located in the Kansai region. The town heavily relied on nuclear-related taxes and subsidies for ¥6.3 billion (US$64.8 million, US$1=¥97.2) of its revenues, or 58 percent of its annual budget in FY2012.[11] When the nuclear crisis shocked Japan's energy policy and the perception of safety surrounding nuclear power plants, the 20-year-old reactors were the center of a debate surrounding possible electric power stability shortages.

Yet, as time progressed and information spread about the possibility of nuclear risk in the form of potentially active fault lines beneath the Ōi Nos. 3–4 reactors and overly theoretical stress tests to assess resistance to natural disasters, actor certainty dissipated in Kansai causing the nature of the debate to shift.

KEPCO argued that its 11 commercial reactors at three separate sites in Fukui (i.e. Mihama, Takahama, and Ōi) accounted for a significant portion of the generation capacity of its vertically integrated and centralized power system in the Kansai region. Nuclear power was 50.9 percent of its total generation mix. In addition, the Ōi reactors themselves were inordinately large and expensive facilities. Reactor No. 3 (1,180 MW) and reactor No. 4 (1,180MW) alone generated on average 15.8 terawatt-hours per annum, or 11 percent of the region's total generation. When both reactors went offline for regularly scheduled maintenance in March and July 2011, respectively, management argued that the prolonged idleness of the two reactors would add considerable risk to the energy security of the region in the approaching summer peak demand months.

Nevertheless, Fukui Governor Issei Nishikawa had indicated that he would not support restarts of idle nuclear reactors until the central government had established new safety standards and demonstrated with objective scientific certainty that the reactors were safe (*Asahi Shimbun*, October 29, 2011: 2). This attitude was shared by Prime Minister Kan who not only insisted on stress tests of the idled reactors using computer models to determine how the plants would perform in the event of an earthquake (phase one), but also insisted on an upgrade of nuclear testing standards commonly used by the European Union in Europe's post-Fukushima stress tests incorporating the lessons learned from the Fukushima crisis (phase two).

On October 28, 2011, stress test results for Ōi No. 3 were quickly submitted to NISA, making them the second reactor to comply with the previous Kan Cabinet's phase-one restart requirements. According to KEPCO, Ōi No. 3 reactor could theoretically withstand an earthquake with 1,260 Gal, 1.8 times larger than the assumed 700 Gal.[12] The reactor and its pipes also could withstand a tsunami 11.4 meters high (37.4 feet), exceeding the assumed height of 2.85 meters (9.35 feet), the report said. What should have satisfied NISA, however, became politically difficult to sell to both the local communities and the national public. Between February 2012 and June 2012, support for restarts had gradually eroded—falling from 41 percent to 30 percent by April 2012. KEPCO, NISA, IAEA, NSC, METI Minister Yukio Edano, Chief Cabinet Secretary Osamu Fujimura, Nuclear Disaster Minister Goshi Hosono, Fukui Governor Issei, host village mayor Hoshitaka Nose, and newly established Prime Minister Yoshihiko Noda himself all signed off on the stress tests results by April 2012 with the intention—to quote the premier—of 'making every effort to improve the safety of those two reactors' and to gain local approval before restarting the Ōi reactors. These restarts, Noda argued, may have echoed KEPCO's focus on protecting the standard of living and stability enjoyed by the Kansai people before the reactors went offline, but whatever economic necessity that might have necessitated the call was crowded out by negative news reports of a possibly active fault line underneath the Ōi reactors coupled with the ongoing crisis at Fukushima Daiichi NPS. Local approval became a politically contentious issue.

After the Kobe Earthquake of 1995 (magnitude 7.3), a minority of scientists became increasingly vocal in their concerns regarding a number of potentially active fault lines running through the region and what an unprecedented earthquake directly underneath or close to one of the reactors would do to the integrity of the infrastructure. These lone voices were largely sidelined in the pre-Fukushima Kansai region in favor of the (then) mainstream view of nuclear safety. KEPCO, complying with the then existing siting regulations, conducted geological surveys of various crush zones in the mid-1980s to determine the status of the faults in question.[13] At that time, nuclear regulations determined an 'active' fault to be any movement in the strata originating up to 50,000 years ago. A special committee was convened to explore the adequacy of those regulations. By 2006, the Cabinet Office's NSC concluded that a more expansive definition of 'active' would be needed. Consequently, the definition expanded from the previous 50,000 years to 120,000 to 130,000 years.

Problems ensued. Perhaps not surprisingly, KEPCO's freshly conducted geological surveys determined that even with the new scientific definition of an 'active' fault originating up to 130,000 years ago, Ōi reactors No. 1 (1979), No. 2 (1979), No. 3 (1991), and No. 4 (1993) were safe. Some scientists disagreed, creating further uncertainty and doubt. In 2005, Katsuhiko Ishibashi, for example, a leading seismologist from Kobe University, and senior advisor to the nuclear industry resigned in protest from the national safety commission's panel likening nuclear reactors in the Kansai region to 'a suicide bomber wearing grenades around his belt' (in Goodspeed 2011: A4). Contradicting KEPCO's geological survey team, Ishibashi insisted that tremors in the Kansai region were stronger than what the reactors were built to withstand. More crucially, the NSC's subsequent regulatory guidelines issued in December 2010 stated that nuclear reactors could not be sited immediately above an active fault, once again calling into question the status of the fault under Ōi power plant.

By the time that METI Minister Kaieda attempted to sell the restarts to the regional governors, municipal officials, and citizens of the Kansai region, the lingering scientific uncertainty regarding active fault lines and their consequences resurfaced as a prominent political issue. Public attitudes demanding nothing less than absolute certainty that the reactors were safe made any tests and assurances almost pointless. NISA insisted that it 'believed appropriate measures [were] currently being taken at reactors 3 and 4 of the Ōi power plant to prevent . . . situations similar to what Fukushima No. 1 plant [was] facing' (Matsutani 2012). Osaka mayor Toru Hashimoto backed by the seven-prefecture Union of Kansai Governments disagreed:

> Stress tests and safety standards are two different things. The only thing that has happened is that a single investigation has approved a set of theoretical numbers based on a theoretical scenario. The commission said nothing about the safety of the plant itself.
>
> (Johnston 2012)

Citing a 2003 report by Kyoto Sangyo University, economics professor Park Sung Joon, who estimated an accident involving Ōi Nos. 3 and 4 could create a radiation leak that would lead to 3.5 million cancer cases in a 50 kilometer radius, including the northern part of the city of Kyoto, anti-nuclear activists argued that the central government and the nuclear regulators were willfully ignoring scientific evidence to support the narrow interests of industry over the wider interests of the Japanese voter.

To be sure, we cannot rule out opportunism in any attempt by local politicians and activists to challenge incumbent actors. What matters in this particular case is how the lack of certainty regarding safety led to a battle over preferences. Without scientific certainty, actor preferences were forced to choose between safety and necessity. Unlike the restart case of Tomari No. 3 in which class-action lawsuits and demonstrations attracted minimal support in Hokkaido in the face of confirmed stress tests and no geological fault lines to consider, the restart case of Ōi Nos. 3 and 4 attracted a growing chorus of local anti-nuclear demonstrations and class-action lawsuits. Each side was aided by select scientific evidence to support their particular issue frame. For pro-restart actors, the various checks and balances afforded the Ōi reactors proved not only their dedication to safety but also the need to protect Kansai from blackouts, transmission grid interruptions, and ever-rising electricity prices. For anti-restart actors, the lack of scientific consensus demonstrated the opposite question: at what point does economic necessity trump the value of human life?

Ultimately, pro-restart supporters won the battle. After months of anti-nuclear rhetoric, the controversial Osaka mayor and neighboring prefectural governors and mayors within the Union of Kansai Governments reversed their positions citing economic concerns, thus suggesting that in times of scientific uncertainty economic considerations are valued highly. Ōi Nos. 3 and 4 were restarted in June 2012.

## Tsuruga No. 2 (scientific authority agrees: unsafe)

If protecting the material self-interest of regional actors via rent-seeking was the explanatory factor in the decision-making process as the regulatory capture model suggests, the puzzling permanent shutdown of Tsuruga No. 2, owned and operated by the Japan Atomic Power Corporation (JAPC), should never have happened. Yet, not only did the consensus of scientific authority find Tsuruga No. 2 to be located upon an active fault in 2012–2013, but also the vast majority of economic interests in favor of its commercial restart backed down quickly in the face of overwhelming scientific agreement. Consequently, this third and final case study emphasizes the power of scientific consensus and certainty to trump economic necessity in times of increased political attention to salient safety issues.

The JAPC, a consortium of Japanese utilities, manufacturers, and other companies, owned and operated four reactors at two separate sites: Tokai Nuclear Power Plant and the Tsuruga Nuclear Power Plant. A wholesale electricity supplier, the JAPC did not service its own region with privately owned transmission and distribution assets, but its exclusive reliance on the wholesale generation of nuclear power to three companies—Chubu EPCO, Hokuriku EPCO, and KEPCO—underscored how critical the stakes had become for its corporate survival in post-Fukushima Japan. Between 1989 and 2005, JAPC profitability slowly deteriorated. Its operating profit margin—the proportion of revenue that remains after deducting costs—fell from roughly 18.5 percent in 1989 to barely 0.9 percent in 2005, making it the least profitable supplier of electricity in Japan (Figure 8.3).[14] Keeping the reactor open would have been crucial for the company to avoid further financial risks.

Tsuruga No. 2 was one of the largest advanced light water reactors of its kind with an installed capacity of 1160 MW. It started its commercial operation on February 17, 1987 after being slated for eight years earlier. Like Tomari No. 3 and Ōi Nos. 3–4 discussed in the previous sections, Tsuruga No. 2 and the JAPC had the strong support of local host politicians, especially Mayor Kazuharu Kawase. Dubbed the 'Ginza of nuclear power plants,' the town of Tsuruga relied heavily on four plants to support the local economy. Approximately 14 percent of the town's population either worked at a nuclear power plant or related facility, or engaged in businesses that were inseparable from nuclear power plants. When their families were included, tens of thousands of people relied on the nuclear power plants to earn a living (*Asahi Shimbun*, May 17, 2013: 14). In addition to investing an estimated

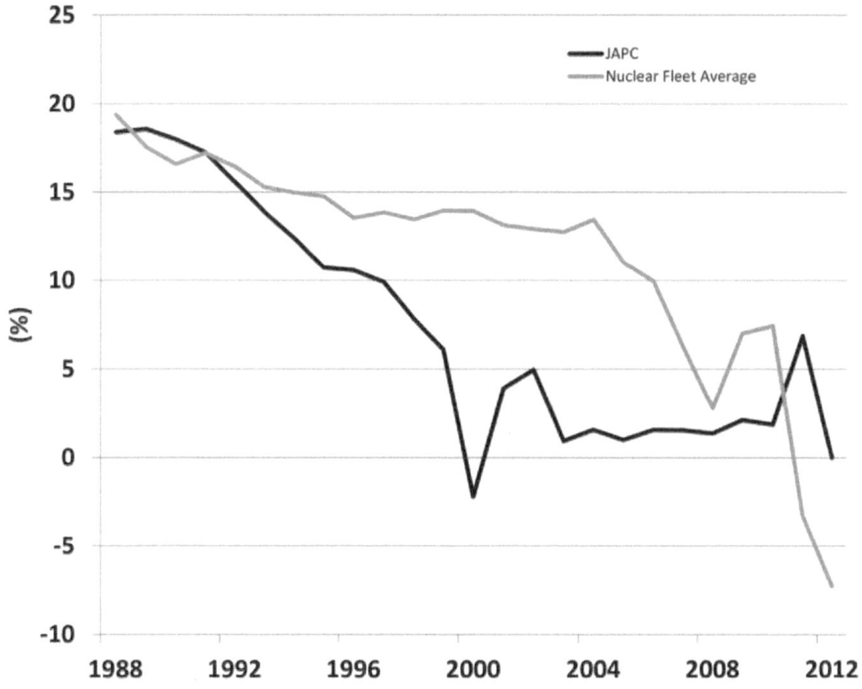


*Figure* 8.3 Operating profit margins of JAPC and total nuclear fleet average (1988–2012)

Notes: On the definition of operating profit and operating profit margin (OPM), see Figure 8.2 notes, supra.

Source: JAPC OPM calculated from financial data culled from *Yūryoku no kōkoku senden-hi [Advertising Spending of Leading Japanese Corporations]* (various years); Nuclear Fleet Average OPM calculated from data culled from Federation of Electric Power Companies financial database (various years).

¥55 billion ($565 million, US$1=¥97.2) in the town from 1970–2006, electric power companies and their nuclear power plants accounted for some 30 percent of the town's annual general budget finances in the form of subsidies and fixed asset taxes on the nuclear power plants (*Asahi Shimbun*, February 3, 2008: 26).

Despite these corporate and local incentives to keep the reactor opened, Tsuruga No. 2 was a casualty of scientific authority. Based on new government guidelines that were revised in September 2006 for the first time in 25 years, the NSC required tougher standards on the nuclear fleet. When Tsuruga No. 2 was inspected, not only did JAPC find that it exceeded the highest number of Gals reported in its previous estimate, but the company acknowledged the existence of a fault line running close to power plant itself. To be sure, this acknowledgement was not the first time that such revelations about Tsuruga No. 2 took place. For years, scientists and media editorials from even the pro-nuclear *Yomiuri Shimbun* had warned that an active fault might lie directly beneath the reactor. The JAPC dismissed such claims, stating that the fault had remained inactive for the past 55,000 years in line with (then) government safety regulations for siting; therefore, the JAPC argued, the company was not guilty of negligence. Yet, Takashi Nakata, a professor at the Hiroshima Institute of Technology, argued that the fault had moved within the last 20,000 years raising serious safety issues even under the pre-existing regulatory framework.

With so much financially on the line for the JAPC without restart of its reactors, the 2011–2013 comparative snapshot of the three incumbent owners and operators of the studied nuclear is instructive. Table 8.1 shows how ineffectual average political resources seem to be in the face of economic necessity. All three companies exhibit similar levels of average cash contributions to Japanese political parties and similar profit margins. Indeed, the JAPC spends far more in advertising, almost as much in political donations to the ruling Liberal Democratic Party (LDP) as HEPCO and KEPCO, and has comparable business assets. Yet, despite these resources, the JAPC proved politically ineffectual against scientific authority.

The creation of the Nuclear Regulatory Authority, or NRA, in September 19, 2012 and its subsequent guidelines on March 19, 2013 changed the dynamic for the JAPC and the company's electric power shareholders. The new authority was neither part of the Cabinet as it was with the NSC nor was it part of the METI as it was with NISA. Despite being under the umbrella of the MOE—a strong proponent of greenhouse gas reductions and increased environmental measures—and therefore predisposed to nuclear power, the NRA was still billed as an 'independent' regulatory body. The NRA consisted of 460 personnel comprising of 350 transferred from NISA, 40 staffed from the NSC, 40 from the Education and Science Ministry, and 30 new hires. Their motto, according to NRA vice-chairman and University of Tokyo emeritus of seismology Kunihiko Shimazaki: 'We must watch nature [natural phenomena] without preconceived ideas'. For Shimazaki, the NRA was to redeem the reputation of seismologists in Japan for having abdicated their responsibilities to let the geological data speak for itself.

*Table 8.1* Average political resources and market shares per company (2011–2013)

| *Company* | *Company's Nuclear Share (%)* | *Regional Market Share (%)* | OPM (%) | *Political Donations (million ¥)* | *Advertising (million ¥)* | *Employees (consol.)* | *Business Assets (million ¥)* |
|---|---|---|---|---|---|---|---|
| HEPCO | 49.4 | 69 | –1.7 | 4,372,010 | 4,177 | 5,457 | 1,765,931 |
| KEPCO | 50.9 | 72 | –3.0 | 13,400,000 | 18,177 | 33,915 | 7,569,823 |
| JAPC | 100 | N/A | 1.9 | 7,931,158 | 72,152 | 2,137 | 868,140 |

Notes: 'Political donations' refers to average cash contributions to the ruling LDP. HEPCO and KEPCO cash contributions are the simple average of the company, individual (read: employee), and subsidiary total contributions to the LDP. JAPC does not directly make cash contributions to the LDP. The number is a calculated weighted average from its joint six utility owners: TEPCO (28.2 percent), KEPCO (18.5 percent), CEPCO (15.1 percent), Hokuriku (13 percent), Tohoku (6.1 percent), and J-POWER (5.3 percent). 'Regional market share' refers the service region of the respective owner and operator of the studied nuclear reactor(s).

Sources: Company's nuclear share, regional market share, and OPM culled from data sourced at and calculated by author from *Denki jigyō binran [The Handbook of Electric Power Industry]*; Political donations culled from data sourced at and calculated by author from *Kanp*; Advertising budgets culled from data sourced at *Yūryoku no kōkoku senden-hi [Advertising Spending of Leading Japanese Corporations]*; JAPC employees and business assets sourced from *Kaisha shiki hō (2014): mijōjō kaisha ban [Corporate quarterly report (2014): unlisted company edition]*; HEPCO and KEPCO employees and business assets sourced from *Kaisha shiki hō (2014) [Corporate quarterly report (2014)]*.

A four-member team consisting of Yasuhiro Suzuki from Nagoya University, Hiroyuki Tsutsumi from Kyoto University, Koichiro Fujimoto from Tokyo Gakugei University, and Takahiro Miyauchi from Chiba University was dispatched to assess Tsuruga No. 2 for two days starting on December 1, 2012. The reactor buildings of JAPC's Tsuruga reactors 1 and 2 were situated approximately 200 meters }(640 feet) from the 25 kilometer (40 mile) long Yanagase-Urazoko active fault, part of which runs through the plant site. However, when the team determined the crush zone (called D-1) running under the Unit 2 reactor building was possibly an active fault that could move in conjunction with the nearby Urazoko active fault, chances of restart were sidelined.

Despite NRA chairman Shunichi Tanaka's insistence that the regulator is 'responsible for confirming whether safety standards are met from a scientific and technological standpoint' and that it 'will not be involved in (a restart decision based on) electricity supply and demand and socioeconomic issues'[15] (*Asahi Shimbun*, October 4, 2012: 1) the May 2013 judgment was a significant financial blow to the company's future and solvency, with decommissioning the reactor estimated to cost in excess of ¥63 billion ($650 million, US$1=¥97.2) alone. For the next two years, JAPC attempted to change the regulator's decision with little political support. In September 2014, the NRA's panel of experts determined that it would not change its 2013 judgment that an active geologic fault existed under the Unit 2 reactor at the Tsuruga nuclear power plant (2014).

With Tokai Power Station decommissioned in 1998, Tsuruga No. 1 forced to decommission because it exceeded the 40-year life expectancy of a nuclear reactor in 2010, and its remaining two reactors left idle in post-Fukushima Japan, the unanimous judgment of the NRA's five-member investigative committee was a strong indication that regulatory capture was not involved in the decision-making process and that the politics of expertise played a crucial role in guiding actor preferences.

## Conclusions

As these three case studies of varying nuclear restarts suggest, scientific ideas surrounding the current state of each reactor seemed to have exercised an independent power to influence the restart prospects by offering persuasive answers to one of the most pressing questions in post-Fukushima Japan: are nuclear reactors safe? This question could be answered in a number of ways, but scientific authority in Japan tackled these questions via 'stress tests' and seismological surveys of potentially active fault lines underneath the reactors that represented the internationally accepted means through which Japanese actors could formulate a political response. It is hardly surprising that the central government in a parliamentary democracy refused to exercise the coercive power of the state to force restarts. Nor is it surprising that supposedly rent-seeking behavior of incumbent power companies was inconsistent and ineffectual, despite the clear financial erosion to all three owners and operators of the nuclear reactors over decades. The implication of these three cases is that the conventional models (null hypotheses) surrounding Japan's regulatory behavior in the government-business relationship are incorrect, or are at least misleading as stated.

In all three cases, the owners and operators of the nuclear reactors faced similar circumstances. All three operators were experiencing declining profitability, declining returns, and rising costs making reactor restarts pressing to the financial health of the company. All three operators had sizeable political resources at their disposal, and all three operators were pitted against NIMBY activists at the local level for years. In addition, these three reactors were located in sympathetic local host communities that relied heavily on the reactors for each town's economic survival. Why, then, were political outcomes so different?

The Tomari No. 3 case in Hokkaido suggests that the power of scientific authority (in this case, stress tests) proved sufficient to curb anti-nuclear demonstrations, class-action lawsuits, and career opportunism of budding Japanese politicians in moments of frenetic uncertainty. Tomari's local host community and Hokkaido governor were staunchly pro-nuclear, to be sure, but the increasingly negative image of the unfolding Fukushima crisis in public opinion could have galvanized anti-nuclear public support in Hokkaido's political marketplace had scientific authority agreed that the reactor was clearly unsafe. It did not. Unlike the next two cases, Tomari No. 3 was not the subject of active fault-line considerations by the scientific community nor was the scientific community divided on the (then) reliability of the stress tests.

The Ōi Nos. 3–4 case in Kansai, by contrast, demonstrates that although causal ideas can be hooks, or justifications for policies adopted for reasons of corporate interest, they can also be pursued even when they exacerbate anti-nuclear sentiment. Like Tomari No. 3, the restarts of Ōi Nos. 3–4 would certainly have been smoother had scientific authority reached a clear consensus on whether the fault lines underneath and near the reactors were active. As experts were divided, so too was Kansai's political marketplace. Actors favoring restarts in the name of economic necessity eventually won the battle in Kansai, but only after extensive wrangling and political posturing over questions of safety versus necessity.

Finally, the Tsuruga No. 2 case stands as a thought-provoking challenge to our null hypothesis surrounding the regulatory capture model and the power of political resources to control energy policy in Japan. If scientists in committee had not unanimously agreed that fault lines beneath Tsuruga No. 2 were active, reactivation might have been a possibility—at least eventually. However, the clear scientific consensus confirming its existence shut down any possibility of a political counter-surge demanding restarts. Despite the clear financial blow to the company and the local host community, Tsuruga No. 2 appears destined for decommissioning with little blowback from pro-nuclear Japanese actors at the time of this writing.

The overall conclusion to be drawn from these cases is that, particularly after an exogenous shock, countries seize upon casual beliefs and pre-existing ideas to serve as cognitive roadmaps that guide them through moments of high uncertainty and doubt. These causal ideas in the form of scientific authority allow them to legitimize and to coordinate their actions. Yet scientific judgment is not always so well heeded. These cases demonstrate that scientific authority is most effective when it contains two attributes: the quality and universality of analysis based on solid empirical research and the practical application of the conclusions. For political economists investigating the government–business relationship in cross-national context, these cases will hopefully serve as much needed food for thought as scholars continue to debate the drivers of institutional change and the slow pace of economic reform in Japan.

## Notes

1 See Article 5: 'The Chairman and the Commissioners of the Nuclear Regulation Authority shall exercise their authority independently.' *Act for Establishment of the Nuclear Regulation Authority* [*Genshiryoku kisei i'inkai setchi hō*] (2012).

2 See the Liberal Democratic Party's 2012 policy pamphlet: Nippon o torimodosu: jyūten seisaku 2012: Jimintō [Taking Back Japan: Important Policy Points of the LDP in 2012].

3 The law has been revised 16 times overall since it was first enacted on October 16, 1972. Following the Fukushima nuclear accident, the law was revised eight times.

4 At least 60 criticality accidents have been recorded since 1945. These have caused at least 21 deaths: seven in the United States, 10 in the Soviet Union, two in Japan, one in Argentina, and one in Yugoslavia. Nine have been due to process accidents, and the others from research reactor accidents.

5 For two examples of this journalistic speculation see: Clenfield and Sato (2007); Onishi and Fackler (2011).

6 Article 18(2) of Section 2 of the *Electric Power Industry Law [Denki jigyō hō]* states that electric power companies shall 'not refuse to supply electricity to meet general demand in its service area . . . without justifiable grounds'.

7 For a sample of works that demonstrate how Japanese regulators were often ineffective, plagued by turf battles and inertia rather than strategic visionaries regulating various sectors for the public good, see Calder (1993); Callon (1995); Friedman (1988); Patrick (1986).

8 For a sample of works that demonstrate how the regulated did not necessarily 'capture' the regulator in the Japanese electric power and energy markets, see Hymans (2011); Johnson (1978); Samuels (1987); Scalise (2009); Scalise (2013).

9 'Stress tests' are defined as targeted reassessments of the safety margins of nuclear power plants. They focus on extraordinary triggering events like earthquakes and floods and the consequences of any other initiating events (e.g. transport accidents, such as airplane crashes) potentially leading to multiple loss of safety functions requiring severe accident management. All the operators of nuclear power plants in Japan had to review the response of their nuclear plants to those extreme situations.

10 Original Japanese: 「不安は多いが、すでに立地し、原発がなくては村はやっていけない」と話す。

11 The four sources of revenue that trickle back to the town are corporate tax on the electric power companies by the central government, fixed asset tax on the nuclear reactors by the town, nuclear fuel tax by the prefecture, and various subsidies.

12 The gal, sometimes called galilleo (symbol Gal), is a unit of acceleration of gravity in the science of gravimetry. It is defined as 1 centimeter per second squared (1 $cm/s^2$).

13 A 'crush zone' is a type of fault that is formed by rocks that become weak after being pressed between strata and turn into clay or pebbles to form a belt-like formation.

14 Japan's average operating profit margin for all retail and wholesale suppliers of electric power in 1989 was 18.9 percent, putting the JAPC slightly below the national average. By 2005, the JAPC was well below the national average of 13.1 percent at 9 percent.

15 Original Japanese: 「規制委は安全基準を満たしているかを確認する立場。電力需要や経済的観点を含めた原発の稼働判断は事業者かエネルギー政策を担当する省庁にしていただくべき。」

## References


2008. '*(Ripoto fukui) shizaisei no ichiyoku ninau jitai: denryoku jigyōsha, tsuruga shi ni 120 oku en kifu/fukui ken* [Fukui Report: Electric Power Companies Donate 12 Billion Yen to Town of Tsuruga: The Role They Played in Town Finances.' *Asahi Shimbun*, February 3.

2011. '*Kōmu, sassoku shidō 3-sen no Takahashi shiji 'hitsuyōijō no jishuku kara dakkyaku o'/Hokkaidō* [Public service: Immediate Action. 'Breaking Away from Self-restraint More than Necessary.' Governor Takahashi on Her Third Hokkaido Election].' *Asahi Shimbun*, April 12.

2011. '*Genpatsu, kataru ka furenu ka, tomari songi iwanai chougi sen ga kokuji: 2011 touitsu chihōsen/Hokkaidō.* [2011 Local Unified Election. Notice for the Election of Tomari Village and Iwanai Village Assembly Members: To Discuss or Not to Discuss Nuclear Power?]' *Asahi Shimbun*, April 20.

2011. '*Genpatsu "dankai-teki haishi" 74-pāsento riyō sansei-ha no 6-wari mo*

*Asahishinbunsha seronchōsa* [Asahi Shimbun Public Opinion Poll: 74 Per Cent of Those Polled Believe that Nuclear Power "Should Gradually Be Shut Down" with 60 Per Cent of Those in Favor of Use Also Favoring Gradual Shut Down].' *Asahi Shimbun*, June 14.

2011. '*Hokuden ni tomari genpatsu no hairo motome, teiso he junbi kai hassoku sapporo de raigetsu shūkai/Hokkaidō* [Hokkaido: Meeting to Convene Next Month in Sapporo to Set Up Arrangment Committee for Lawsuit against Hokkaido Electric Power Co. Suing for the Decommissioning of the Tomari Nuclear Reactor].' *Asahi Shimbun*, June 20.

2011. '*Tomari genpatsu saikadō "mitomenai de" Hokkaidō no 38-jin teiso* [38 People Sue in Hokkaido for "Not Recognizing" the Tomari No. 3 Nuclear Restart].' *Asahi Shinbun*, August 2.

2011. '*Genpatsu anzen, mazu jiko saiten hōkoku-sho "yoyū mashita" Ōi 3-gōki, taisei hyōka* [Nuclear Power Plant Safety: First Self-Assessment Stress Test Report Towards "Increased Allowance" of Ōi Reactor No. 3].' *Asahi Shimbun*, October 29.

2011. 'National Diet of Japan Fukushima Nuclear Accident Independent Investigation Commission. [*Tokyō denryoku fukushima genshiryoku hatsudensho jiko chōsa i'inkai hō*].' In House of Representatives Steering Committee Chairman. Ed. 112, National Diet, Tokyo.

2012. '*Genpatsu, chū ni uku saikadō handan: kisei'i ha sezu* [With Government Wary, Decision on Reactor Restarts Up in the Air].' *Asahi Shimbun*, October 4.

2012a. 'Act for Establishment of the Nuclear Regulation Authority [*Genshiryoku kisei i'inkai setchi hō*].' *Act No. 47*. Japan.

2012b. '*Nippon o torimodosu: jyūten seisaku 2012: jimintō* [Taking Back Japan: Important Policy Points of the LDP in 2012].' LDP Headquarters, Tokyo.

2013. '*(Shasetsu) Datsugenpatsu to jimoto, tsuruga o moderu kesu ni* [Editorial: Tsuruga Should Be Model for Breaking Free of Nuclear Energy].' *Asahi Shimbun*, May 17.

2014. 'Regulatory Panel in Japan Affirms Decision That Will Force Decommissioning of Tsuruga Reactor.' *Energy Monitor Worldwide*, September 6.

Aldrich, D.P. 2008. *Site Fights: Divisive Facilities and Civil Society in Japan and the West*. Ithaca, NY: Cornell University Press.

Bagchi, A.K. 2004. *The Developmental State in History and in the Twentieth Century*. New Delhi: Regency Publications.

Baumgartner, F.R., Berry, J.M., Hojnacki, M., Kimball, D.C. and Leech, B.L. 2009. *Lobbying and Policy Change: Who Wins, Who Loses, and Why*. Chicago: University of Chicago Press.

Broadbent, J. 1998. *Evironmental Politics in Japan: Networks of Power and Protest*. Cambridge: Cambridge University Press.

Byrne, J. and Hoffman, S.M. 1996. *Governing the Atom: The Politics of Risk*. New Brunswick, NJ: Transaction Publishers.

Calder, K.E. 1993. *Strategic Capitalism: Private Business and Public Purpose in Japanese Industrial Finance*. Princeton, NJ: Princeton University Press.

Callon, S. 1995. *Divided Sun: MITI and the Breakdown of Japanese High Tech Industrial Policy, 1975–1993*. Stanford, CA: Stanford University Press.

Camerer, C.F. 1998. 'Behavioral Economics and Nonrational Organizational Decision Making.' In *Debating Rationality: Nonrational Aspects of Organizational Decision Making*. J.J. Halpern and R.N. Stern (eds). Ithaca, NY: Cornell University Press, 53–77.

Camerer, C.F. and Thales, R.F. 1995. 'Ultimatums, Dictators, and Manners.' *Journal of Economic Perspectives*, 9(2): 209–19.

Campbell, J.C. 1992. *How Policies Change: The Japanese Government and the Aging Society*. Princeton, NJ: Princeton University Press.

Chappell, H.W. 1982. 'Campaign Contributions and Congressional Voting: A Simultaneous Probit-Tobit Model.' *Review of Economics and Statistics*, 64(1): 77–83.

Clenfield, J. and Sato, S. 2007. 'Japan Nuclear Energy Drive Compromised by Conflicts of Interest.' *Bloomberg News*, December 12.

Dauvergne, P. 1993. 'Nuclear Power Development in Japan: "Outside Forces" and the Politics of Reciprocal Consent.' *Asian Survey*, 33(6): 576–91.

Derthick, M. and Quirk, P.J. 1985. *The Politics of Deregulation*. Washington, DC: Brookings Institution.

Ferguson, C.D. and Jansson, M. 2013. 'Regulating Japanese Nuclear Power in the Wake of the Fukushima Daiichi Accident.' FAS Issue Brief. Federation of American Scientists, May.

Friedman, D. 1988. *The Misunderstood Miracle: Industrial Development and Political Change in Japan*. Ithaca, NY: Cornell University Press.

Friedman, M. 1993. 'George Stigler: A Personal Reminiscence.' *Journal of Political Economy*, 101(5): 768–73.

Goldstein, J. and Keohane, R. 1993. *Ideas and Foreign Policy: Beliefs, Institutions and Political Change*. Ithaca, NY: Cornell University Press.

Goodspeed, P. 2011. 'Hubris Led Us Toward Nuclear Danger; Warnings Hidden by Years of Cover-Ups.' *National Post (Canada)*, March 15.

Greene, J.M. 2008. *The Origins of the Developmental State in Taiwan: Science Policy and the Quest for Modernization*. Cambridge, MA: Harvard University Press.

Grenzke, J.M. 1989. 'Pacs and the Congressional Supermarket: The Currency Is Complex.' *American Journal of Political Science*, 33(1): 1–24.

Hommerich, C. 2012. 'Trust and Subjective Well-Being after the Great East Japan Earthquake, Tsunami and Nuclear Meltdown: Preliminary Results.' *International Journal of Japanese Sociology*, 21(1): 46–64.

Horiuchi, Y. 2009. 'Understanding Japanese Politics from a Local Perspective.' *International Political Science Review*, 30(5): 565–73.

Hymans, J.E.C. 2011. 'Veto Players, Nuclear Energy, and Nonproliferation: Domestic Institutional Barriers to a Japanese Bomb.' *International Security*, 36(2): 154–89.

Johnson, C. 1978. *Japan's Public Policy Companies*. Washington, DC: American Enterprise Institute for Public Policy Research.

Johnson, C. 1982. *Miti and the Japanese Miracle: The Growth of Industrial Policy, 1925–1975*. Stanford, CA: Stanford University Press.

Johnson, C. 1999. 'The Developmental State: Odyssey of a Concept.' In *The Developmental State*. M. Woo-Cumings (ed.). Ithaca, NY: Cornell University Press, 32–60.

Johnston, E. 2012. 'Kansai Politicians Want New Safety Regime Before Oi Reactor Restart.' *The Japan Times*, March 28.

Jones, B.D. 1994. *Reconceiving Decision-Making in Democratic Politics: Attention, Choice, and Public Policy*. Chicago: The University of Chicago Press.

Jones, B.D. 2001. *Politics and the Architecture of Choice: Bounded Rationality and Governance*. Chicago: The University of Chicago Press.

Kingdon, J.W. 1984. *Agendas, Alternatives, and Public Policies*. Boston, MA: Little, Brown.

Kingston, J. 2012. 'Japan's Nuclear Village.' *The Asia-Pacific Journal*, 10(37).

Kodderitzsche, L. 1991. 'Japan's New Administrative Procedure Law: Reasons for its Enactment and Likely Implications.' *Law in Japan*, 24: 114–15.

Kurokawa, K., Ishibashi, K., Ohima, K., Sakiyama, H., Sakurai, M., Tanaka, K., Tanaka, M., Nomura, S., Hachisuka, R. and Yokoyama, Y. 2012. 'The Official Report of the Fukushima Nuclear Accident Independent Investigation Commission.' Tokyo, JAPAN T.N.D.O.

Kushida, K.E. 2012. 'Japan's Fukushima Nuclear Disaster: Narrative, Analysis, and Recommendations.' Shorenstein APARC Working Paper, Stanford University Shorenstein Asia-Pacific Research Center, December 15.

Lesbirel, S.H. 1998. *Nimby Politics in Japan: Energy Siting and the Management of Environmental Conflict*. Ithaca, NY: Cornell University Press.

Maclachlan, A. 2011. 'IAEA Chief Proposes Universal Nuclear Plant Safety Reviews.' *Platts Nucleonics Week*, 52(25).

March, J. 1994. *A Primer on Decision-Making*. New York: Free Press.

Marshall, J.M. and Navarro, P. 1991. 'Costs of Nuclear Power Plant Construction: Theory and New Evidence.' *The Rand Journal of Economics*, 22(1): 148–54.

Marx, K. 1970. *Critique of Hegel's 'Philosophy of Right'*. Cambridge: Cambridge University Press.

Matsutani, M. 'Oi Reactor Stress Tests Approved by NISA.' *The Japan Times*, February 14.

Mayer, K.R. 1991. *The Political Economy of Defense Contracting*. New Haven, NJ: Yale University Press.

McLaughlin, T.P., Monahan, S.P., Pruvost, N.L., Frolov, V.V., Ryazanov, B.G. and Sviridov, V.I. 2000. 'A Review of Criticality Accidents.' Report no. LA–13638. Los Alamos National Laboratory. New Mexico, May. Retrieved at: www.osti.gov/scitech/servlets/purl/758324.

Mulgan, A.G. 2002. *Japan's Failed Revolution: Koizumi and the Politics of Economic Reform*. Canberra: Asia Pacific Press.

Onishi, N. and Fackler, M. 2011. 'Japanese Officials Ignored or Concealed Dangers.' *The New York Times*, May 17.

Patrick, H. 1986. *Japan's High Technology Industries: Lessons and Limitations of Industrial Policy*. Seattle, WA: University of Washington Press.

Peltzman, S. 1993. 'George Stigler's Contribution to the Economic Analysis of Regulation.' *Journal of Political Economy*, 101(5): 818–32.
Posner, R.A. 1974. 'Theories of Economic Regulation.' *The Bell Journal of Economics and Management Science*, 5(2): 335–58.
Priest, G.L. 1993. 'The Origins of Utility Regulation and the "Theories of Regulation" Debate.' *Journal of Law and Economics*, 36(1): 289–323.
Quirk, P.J. 1988. 'In Defense of the Politics of Ideas.' *The Journal of Politics*, 50(1): 31–41.
Sakai, T. 2012. 'Fair Waste Pricing: An Axiomatic Analysis to the Nimby Problem.' *Economic Theory*, 50(2): 499–521.
Samuels, R.J. 1987. *The Business of the Japanese State: Energy Markets in Comparative and Historical Perspective*. Ithaca, NY: Cornell University Press.
Scalise, P.J. 2009. 'The Politics of Restructuring: Agendas and Uncertainty in Japan's Electricity Deregulation.' D.Phil. doctoral dissertation, University of Oxford.
Scalise, P.J. 2013. 'Who Controls Whom? Constraints, Challenges, and Rival Policy Images in Japan's Post-War Energy Restructuring.' In *Critical Issues in Contemporary Japan*. J. Kingston (ed.). London: Routledge, 92–106.
Schwartz, F.J. 2001. *Advice and Consent: The Politics of Consultation in Japan*. Cambridge: Cambridge University Press.
Sorauf, F.J. 1992. *Inside Campaign Finance: Myths and Realities*. New Haven, NJ: Yale University Press.
Stigler, G.J. 1971. 'The Theory of Economic Regulation.' *Bell Journal of Economics and Management Science*, 2(1): 3–21.
Stigler, G.J. and Friedland, C. 1962. 'What Can Regulators Regulate? The Case of Electricity.' *Journal of Law and Economics*, 5: 1–16.
Thaler, R.H. 1991. *Quasi Rational Economics*. New York: Russell Sage.
Thaler, R.H. 1992. *The Winner's Curse: Paradoxes and Anomalies of Economic Life*. Princeton, NJ: Princeton University Press.
Von Hippel, F.M. 2011. 'It Could Happen Here.' *New York Times*, March 24.
Welch, W. P. 1982. 'Campaign Contributions and Legistlative Voting: Milk Money and Dairy Price Supports.' *Western Political Quarterly*, 35(4): 478–95.
Woll, C. 2008. *Firm Interests: How Governments Shape Business Lobbying on Global Trade*. Ithaca, NY: Cornell University Press.
Woo-Cumings, M. 1999. *The Developmental State*. Ithaca, NY: Cornell University Press.

# 9 Drifting to new worlds

## On politics and science in modern biotechnology

*Haig Patapan*

Jonathan Swift's *Gulliver's Travels* tells the story of the flying island of Laputa populated by experts in mathematics, music, astronomy and technology who keep their island airborne by means of magnetic levitation. Absorbed in listening to the music of the spheres and contemplating the stars, the Laputans seem less able to address their immediate practical needs; not only are they ill dressed and poorly housed (because they do not see the 'human' and therefore use compasses and quadrants instead of tape measures), but they are forced to exercise a form of tyrannical rule over the Balnirabi, the land over which the Laputa hovers. Swift's amusing account of Laputa, recalling Aristophanes' *The Clouds* where Socrates the natural scientist is suspended mid-air in a 'think tank', is a modern reminder of the tension, even contest that has always existed between the intoxicating charms of philosophical contemplation and the immediate and practical demands of political life. Socrates' execution by the Athenians for impiety and corrupting the young is one of the most famous and earliest instances testifying to the seriousness of such a contest. Of course, Socrates is not unique. Before him we have the case of Anaxagoras, the cosmological scientist who survived the charge of impiety only through the intervention of his famous pupil, the Athenian statesman Pericles. After Socrates' death, we have Plato's famous student Aristotle fleeing Athens, famously claiming that he would not let the Athenians sin twice against philosophy. These earliest instances of the tension between philosophy and politics would soon be transformed by the increasing authority of revealed religion, especially Christianity and Islam, yielding new struggles and ambitious attempts to resolve them by the Enlightenment and modern science.

Two powerful impulses asserted themselves in this struggle between philosophy, which soon became 'natural' science or more simply 'science', and politics. The first was the dominance of politics (and especially religion) over science, deploying all aspects of political force, from censorship to criminal and capital punishment, to assert its authority and to protect itself from what it saw as the dangers of impiety, human hubris or overreach. The requirement that science should conform to various forms of political ideology,[1] and more subtly the claim that science is fundamentally indistinguishable from politics[2] are but recent versions of such attempts at political dominance over science. Countering such an impulse was the attempt to transform politics, to 'enlighten' it by introducing scientific knowledge and principles into political practice to refound it on a more rational, and peaceful basis. Bacon's *Great Instauration* (1620), Hobbes' *Leviathan* (1654) with its geometric political science, the *philosophes' encyclopaedias*, Bentham's *An Introduction to the Principles of Morals and Legislation* (1789) as a scientific approach to law and morality, Condorcet's 'empire of reason', and his disciple Auguste Comte's proposal for a 'positive philosophy', a new 'sociology' to match the advances in the natural sciences are indicative of the various attempts to refashion politics in the image of science.

Mediating between these two impulses was the view that politics and science could coexist, albeit in an unequal partnership. First developed by Augustine, who attempted to reconcile Platonic philosophy with Christian theology and subsequently taken up by Thomas Aquinas, who sought to do the same with Aristotelian political thought, this formulation sought to demarcate separate spheres of influence between faith and reason. Reason would have its autonomy and authority recognized, but only within the purview of theology. Beyond its sphere unaided reason could not proceed and had to defer to theology. Therefore philosophy was to be the 'handmaid' of theology, complementing faith and piety.[3] Drawing upon this resolution, science as handmaid to politics has become an influential way of reconciling the two powerful impulses we have noted above.[4] It is certainly the foundational model for modern research, with scientific projects funded by governments on the basis of specific benefits to national interest. But does such a mutual accommodation work? Does politics allow science the freedom it has always demanded to pursue its questions? Does science limit its insights to the politically expeditious, confining its role to that of the 'handmaid' and not mistress?

In this chapter I examine these questions and how the tension between science and politics manifests itself in practice by focusing on the case of biotechnology. Biotechnology is an especially useful case study for a number of related reasons. Biotechnology, along with computing, nanotechnology and particle physics, is a new science with revolutionary potential. Yet, unlike these other disciplines, biotechnology and especially genetics seems to go beyond the therapeutic and remedial to challenge our deep-seated conceptions of what life is and what it means to be human. Its revolutionary research therefore also issues a provocative challenge to politics. In addition to its significance, biotechnology is an especially revealing case for understanding the interaction between political and scientific authority because it is located at the intersection or confluence of science (including a number of new sub-disciplines), technology, politics and commerce.

In the first part of the chapter I briefly outline the history of biotechnology, from its early origins as zymotechnology to 'new' biotechnology. This overview, in addition to showing what is new in biotechnology, will form the basis for our examination of how science and politics interact in practice in this domain. As we will see, the history of biotechnology reveals a dynamic interchange between science and politics, where scientific innovation is funded and supported by politics and shaped by national and commercial interests. Though science is an important partner in this relationship, what we discern from this account is the overarching dominance of politics and commerce in directing scientific innovation.

It is this view of political mastery over science that is questioned in the second part of the chapter, which begins with a close analysis of major biotechnological innovations such as assisted reproductive technologies, genetically modified foods, cloning and stem cell research. This discussion shows how innovations in genetics and the great promises they hold for human betterment allow scientific innovation a powerful political influence. But this political influence, both in form and extent, is more subtle, complex and profound. The way influence is wielded is unusual to the extent that it is ostensibly exercised without intention and strategy, in the form of an ever-increasing number of innovations seemingly indifferent to political and social consequences. Yet to the extent that these innovations question prevailing political conceptions of human life, the importance of the individual and the role of family, they incrementally and profoundly challenge not only political authority, but the foundations of politics itself. Often conceptualised or posed as problems of 'ethics', these confrontations between scientific insights and political authority show how science with its innovations redefines and reconceives the language and substance of politics. These confrontations also show how the influence of the 'useful' or utilitarian considerations in contemporary politics, combined with the industrial and multinational character of biotechnology means that politics is constrained in countering the innovative force of science. The core argument of the chapter therefore is that the seemingly authoritative influence of politics conceals a more powerful scientific impulse that seems impervious to political control, yet does not appear to be guided by any overarching motive, purpose or strategy of its own, incidentally refashioning and reshaping the contours and very foundations of politics.

## Biotechnology and politics

Modern biotechnology, one of the most advanced and potentially revolutionary fields of science, can trace its origins to beer.[5] One of the earliest recorded instances of biotechnology is the ancient practice of using barley, the first cultured cereal, for brewing and beer production. Indeed, it has been the major products of fermentation—alcohol, beer and wine—that have driven the scientific, technical and economic interest in fermentation research. Most of this research has historically been based on trial and error, with experimentation generally uninformed by theoretical considerations. The major changes to this approach took place with Pasteur's 1854 discovery that fermentation could be traced to a microbe, creating the new science of microbiology and providing important new insights into better brewing practices with significant economic consequences.[6] This microbial research resulted in the first use of antiseptics in surgery in 1865, and vaccines for anthrax and rabies in 1881. These discoveries in turn laid the groundwork for further developments from 1890 to 1950 with the discovery of biochemistry by Fischer and Buchner in the 1890s and Fleming's findings of the anti-bacterial properties of the fungus *Penicillium notatum* in 1928.[7] The turning point for penicillin came with the investigations by Florey, Heatley and Chain at Oxford University and the subsequent involvement of the US government and the four companies Merck, Squibb, Pfizer and Lederle in developing sufficient stocks of the drug for the US Army's invasion of Europe in the spring of 1944.[8] Just as technical innovation in fermentation was driven by the First World War requirements for acetone and butanol for explosives production, the Second World War's demand for penicillin laid the foundations for future research on antibiotics.

It is in the period from the 1950s to the present that we see the transformation of traditional or classical biotechnology to what we would consider the modern biotechnology of genetic innovation. By the 1950s the advances in fermentation technology had produced new antibiotics (such as streptomycin), amino acids, organic acids, carbohydrates, vitamins, solvents and enzymes. But research in biotechnology in the 1960s and 1970s had diversified into a number of disciplines, ranging from molecular biology and genetics, to microbiology, cell biology, biochemistry and chemical engineering, each employing different theoretical approaches, working procedures, languages and analytical tools. This situation changed in the 1970s and 1980s when 'biotechnology' gained the attention of government agencies in Germany, UK, Japan and the US as a field of innovative potential and economic growth. The 1974 Dechema Report issued by a German chemical technology organization for the German Ministry for Education and Science was the first systematic proposal for biotechnology funding and development.[9] Noting the conservatism of German and European chemical, pharmaceutical and food industries regarding biotechnology, the Report emphasized the need for interdisciplinary work as well as cooperation of applied research and industry. As a consequence, the German government began supporting biotechnology research and development to encourage industry involvement, an example of political, commercial and scientific collaboration. Similar studies were made in the UK, Japan and France.[10] Yet, as these studies revealed, the emphasis remained on classical biotechnological innovations. This was to change fundamentally in the 1980s. The final breakthrough in the molecular basis of genetics is credited to the Watson and Crick publication of the DNA structure in 1953. But what came to be known as the DNA revolution was made possible by numerous specific discoveries, such as investigations into restriction enzymes, phage (bacterial virus) genetics and the study of plasmids.[11] One of the most important of these was the Berg, Cohen and Boyer's 1972 recombinant DNA (r-DNA) technology constructing the first recombinant plasmids and introducing them into bacteria able to retain genetically modified plasmids while growing.[12] Equally important, however, was the economic development of these discoveries, through new biotechnology companies such as Cetus (1971), Genex (1977), Amgen and Biogen (1980), Genzyme and Chiron (1981) and Genecor (1982).[13] Industrial development of r-DNA technology resulted in the production of drugs for pharmaceutical and medical use.

The significance of these innovations for 'new' biotechnology is evident from a US Office of Technology Assessment (OTA) study in 1984.[14] DNA discoveries meant that the OTA study now emphasized genetic engineering and rDNA technology, and their potential commercial exploitation of r-DNA, cell fusion and novel bioprocessing techniques. Recognizing the comparative advantage of the US to its major potential competitors, such as Japan, Germany, the UK, Switzerland and France, the report recommended government funding of basic and applied research, in addition to financing and tax incentives for firms.

The OTA report had noted the importance of genetics in transforming the 'old' biotechnology into new biotechnology, with new opportunities for commercialization. This assessment proved to be accurate in the subsequent development of genetic research and recombinant technologies. The first recombinant industrial product was human insulin developed by Genentech in cooperation with Eli Lilly in 1978.[15] This innovation proved to be a turning point in industry perceptions of genetic engineering. Venture capitalists and transnational corporations soon transformed genetic engineering from an area of academic research to a field of industrial technology, encouraging scientists to become equity owners, corporate executives, members of scientific advisory boards and industry consultants. This trend was aided by the 1980 US Supreme Court decision in *Diamond v Chakrabarty*, which ruled narrowly (5–4) that a patent could be obtained for a genetically engineered bacterium because it was a 'non-naturally occurring product of human ingenuity' (Wright and Wallace 2000: 46–7).[16] Soon other r-DNA products followed, including human growth hormone in 1983, interferon and hepatitis B vaccine in 1986, tissue plasminogen activator in 1987 and erythropoietin in 1989. Recombinant technologies and the potential of metabolic engineering transformed biotechnology and led to its extraordinary growth. As an indication of this transformation, patents represent an important benchmark in the development of new biotechnological medicine. In 1978, 30 biotechnology patents were requested; by 1988 the figure had reached 500. In contrast, between 1994 and 1997, the biotechnology industry entered over 48,000 (Wright and Wallace 2000: 48). More recent developments in the evolution of biotechnology include Dolly the female domestic sheep, the first mammal to be cloned from an adult somatic cell using the process of nuclear transfer (Maienschein 2001), and the Human Genome project initiated in 1990 and completed in 2003, mapping all the genes in the human DNA.[17]

This account of biotechnology shows the close links between politics, science and commerce. The influence of politics is evident in the way it has sponsored and encouraged the industry, from its earliest beginnings in beer, wine and food production, to the demand of antibiotics for the war effort, and more recently in government support for developing 'new' biotechnology. Political endorsement of biotechnology has rarely been for knowledge as such; rather it has been for its contribution to public welfare, national advantage and as part of its commercial or industrial policy. In this context we can see in the historical development of biotechnology the influence of each nation's unique cultural and historical character. Equally important in this account has been the active involvement of science and scientists in the evolution of biotechnology. Science has collaborated with politics and industry to develop new insights and techniques that have greatly benefited both politics and commerce. What is notable in the case of biotechnology is that where the impulse of science has been to fragment and lead to diversity in disciplines, it has been politics and industry funding and support that has forced interdisciplinary focus and in effect made possible the 'new' biotechnology. Equally, scientists have taken an active part not only in the development of biotechnological innovations, but also in their commercialization, with prominent scientists becoming major stockholders in companies that have exploited scientific breakthroughs, underscoring the powerful influence of commerce and industry in the development of biotechnology, from its very beginnings in brewing to the modern influence of multinational corporations.[18]

This brief account of the development of biotechnology, from practical initiatives in better brewing techniques and food preservation, to microbiology, and finally to modern biotechnology shows the subtle relationship between science (and its various disciplines), commerce and industry and politics shaped by national interest. It shows how each of these approaches—the scientific, commercial and political—differs in important respects from the others, yet these differences are not so defined or pronounced that there is an unbreachable impasse between them. Indeed, the evolution of biotechnology testifies to the contrary, where science, industry and politics, opportunistically or strategically, are willing to benefit through mutual accommodation and compromise. But such accommodation should not take us to the other extreme, the view that each is equally authoritative. What is clear from the development of biotechnology in general is the dominance of the political over the scientific. We saw this in the initiative governments took in funding modern biotechnology, in effect creating a new science discipline, in the decisive role played by the judiciary in determining who owns scientific innovations and discoveries, and perhaps most significantly, in determining the practical application of innovations, in some cases imposing complete limits on scientific research. Mutual accommodation therefore co-exists with a ranking or hierarchy where politics predominates both through financial and legislative authority to guide, direct and in some cases prohibit scientific authority. It would seem, therefore, that though science as handmaiden theory may not do justice to the nuanced interplay between science and politics, it is in substance a correct assessment of the relative authority of the two.[19]

## Biotechnological innovation and the transformation of politics

Yet the apparent primacy of political authority over science tends to obscure the profound and subtle way biotechnology transforms politics. It is perhaps easy to miss this influence because it manifests itself in different political aspects and expressions, such as the problem of ethics, or corporate dominance, or globalization.[20] These themes undoubtedly reveal a tension between the advance of science and the challenge it presents to politics. But in their various formulations they do not show with sufficient clarity the subtle force and reach of the scientific transformation of politics. To see the nature of these biotechnological innovations and the challenges they present to politics it is instructive to examine three major policy areas—Assisted Reproductive Technology; genetically modified food; cloning and stem cell research.[21] Though each policy area is characterized by a subtly different interaction between science and politics, our examination will show a common tendency of biotechnology to transform the language and substance of politics in fundamental respects.

Assisted Reproductive Technology (ART) was developed to assist those who had fertility problems, defined as 'not being able to get pregnant after trying for one year' (West 2007: 32). The causes of infertility range from poverty, poor levels of medical treatment and high levels of sexually transmitted diseases in poorer countries, to delay in childbearing in modern developed nations. Treatments range from injecting sperm into an egg, transferring a donor egg, embryo freezing, assisted hatching, sperm storage, artificial insemination and drug therapies to stimulate egg production. Because in-vitro techniques are expensive and time-consuming, the vast majority of women who have fertility problems do not seek treatment. The first and most famous example of a 'test tube baby' was British born Louise Brown, delivered by caesarean section on July 25, 1978. Since this pioneering effort ART has developed into an international, multi-billion dollar industry that has helped thousands of infertile mothers conceive and bear children. The overwhelming public popularity of ART and the relatively unregulated nature of these procedures have raised concerns regarding the medical licensing and costs of ART. But the more profound question concerns the way that ART has also potentially altered politics by compelling it to respond, through laws or litigation, to the historically unique possibilities now made feasible by the new technology. These range from the questions regarding ownership of frozen embryos, how long they should be kept and what happens in the case of divorce or death, to the legal roles and responsibilities of natural and surrogate mothers and fathers. Though enhanced fertility technology clearly endorses and strengthens the concept of the family, it also subtly and incrementally challenges political leaders to reconsider the meaning of 'family', who is a father, mother and child, and whether an embryo is private property or warrants protection as a human being.[22]

Genetically Modified (GM) foods have been a major growth area in biotechnology. Traditionally, new plant varieties could only be developed through a very slow process of cross-pollination with new varieties confined to the same species. r-DNA technology now allows the transfer of one or more genes from one organism to another, crossing biological species and increasing the rate of crop modification to enhance their shelf-life and resistance to diseases and pests. GM has been so successful that about 10 per cent of the world's seed supply now includes bio-crops, with about half of the world's agribusiness controlled by the 'Big Three' biocorp companies of Monsanto, Syngenta, and Aventis.[23] The immediate challenge to GM food technology concerns the practical matters of food safety. But the ability of GM technology to cross species boundaries also poses larger and more profound political questions. Beyond genetic modification such as Calgene's 'Flavr Savr' tomato with improved shelf-life (one of the first GM products approved by the US Department of Agriculture in 1992), there is the prospect of cross-species experimentation. In 1984 scientists produced a 'geep', a combination of goat and sheep embryos and more recently quail brain has been combined with chickens to produce baby chicks that sound like quails while pigs have been cloned to generate their own omega-3 fatty acids usually found in oily fish. Though still at an early stage, these experiments show how the adoption of GM crops was the first incursion into a radically new world of apparently unlimited cross-species innovation, with the extreme instance being 'chimeras', genetic cross-breeds based on the living cells of human beings and other animals. The first human chimera was produced in 2003 when researchers in China fused genetic material from humans and rabbits. In 2005 American researchers combined human and cat protein to treat human allergies to felines. Stanford researchers have raised mice with brains that have traces of human brain cells. It is not so much that these experiments raise significant moral and ethical questions concerning the boundaries between species and more specifically, the difference between humans and animals. More fundamentally, it is the intrusion into politics about a new way of thinking about living things. 'There aren't', according to ethicist Henry Greeley, '"human genes" and "goat genes". There are just genes that have different functions' (cited in West 2007: 104). To think in terms of 'gene functions' rather than 'humans' and 'goats' represents a fundamental change in political thinking and therefore scientific transformation of politics. Incremental and incidental, these transformations will in due course compel politics to confront the historically unique and unprecedented situation of being forced to determine what was once outside ordinary political adjudication: the proper boundaries between living and non-living, human and non-human.[24]

These questions are being posed even more directly in the context of cloning and stem cell technology (Capps 2008; Caulfield 2003). Cloning involves the somatic cell nuclear transfer, where the nucleus of an unfertilised egg cell is removed and replaced with material from the nucleus of a somatic cell stimulating it to divide, and embryo splitting, the artificial division of a cell into two or more cells, with the resulting embryo placed in a female womb to produce identical offspring. Since Ian Wilmut of the Roslin Institute in Edinburgh announced the successful cloning of Dolly the sheep, there have been clones of rabbits, calves, pigs and monkeys. This biotechnology has opened the possibility of reproductive and non-reproductive (or therapeutic) cloning. Reproductive cloning aims to produce a genetically identical organism and placing it in a womb for the purpose of reproduction. Reproductive cloning is contentious because unlike ART it raises the possibility of asexual reproduction of identical human beings. Therapeutic cloning seeks to provide compatible tissues and organs for replacement therapy. A survey of 30 developed nations around the world found seven had no national legislation overseeing any kind of cloning, all the rest banned reproductive cloning, while 17 nations ban therapeutic cloning. Countries such as Belgium, China, Finland, Greece, India, Israel, Korea, New Zealand and the United States have no national legislation on reproductive cloning and allow therapeutic cloning (West 2007: 71; Capps and Campbell 2010; Isasi and Knoppers 2006; Montpetit *et al.* 2007). Stem cell research has been as contentious as human cloning. Traditional stem cell therapy, such as transplanting adult bone-marrow cells into patients for the treatment of leukaemia, was limited by the need to match donor and host to avoid rejection and the limitation of adult cells to a specific function. In 1998 James Thomson at the University of Wisconsin identified human embryonic stem cells that could grow into many different lines, forming the building blocks for the 210 different kinds of tissues in the human body. Removed from foetuses or placenta and umbilical cords, these cells can be cultured into skin, blood, heart or other human tissues and used for medical treatments. The 'pluripotent' cells that can be used as generic stem cell lines to generate new tissues and organs need to be removed between 9 and 11 days from the blastocyst, the 200 cells that form about nine days after conception, before these cells receive chemical signals to specialise. Stem cell technology, with the potential to repair or regenerate cells appears to offer extraordinary therapeutic possibilities, ranging from spinal cord repair, to treating diabetes and multiple sclerosis. The problem with this technology and the major reason it has been controversial is that it involves the use of cells from terminated pregnancies and the destruction of embryos. The biotechnology of cloning and stem cells has been politically contentious because it poses the most disturbing questions. Stem cell research forces us to ask when life begins and whether all living things, especially other human beings, are for our use. Human cloning presents arguably the most radical challenge to politics, potentially overturning all laws based on individual ownership, intentionality and culpability. In the promise of asexual reproduction it potentially destroys the concept of the family and all those institutions and traditions associated with it. Even more provocatively, cloning seems to challenge our bedrock political notion of time and mortality, and a political world that presumes generational change and therefore confronts and negotiates the debts owed to the past and the future in the form of intergenerational justice.[25]

Our general overview of the major biotechnological innovations of ART, GM foods, stem cell and cloning show how major advances in biotechnology have important consequences for politics. What we see in this account are new insights, discoveries and techniques that present fundamental challenges to ideas and institutions that are at the core of all political communities. These challenges go beyond perennial political questions, such as who is a citizen, who should rule, what is the best political arrangement, to questions that were once considered the province of abstract philosophical or metaphysical debates but are now important questions of public policy and legislation. Consider some of the questions that politics has been forced to confront and determine as a consequence of biotechnological innovation. Can we now confidently distinguish between what is living and what is not? What is 'human' and when do things 'become' human? Is it possible to demarcate between human and non-human and if not, on what basis is one to rank over the other? If indeed a world seen through the lens of genes reveals the inherent plasticity of all things, does this not mean that all those structures and institutions that presently shape politics are now at best questionable? More specifically, does not biotechnological innovation raise radical questions not just about the political meaning of individuality and family but the very existence of such things, thereby questioning not just those provisions such as criminal laws, property ownership and constitutive enactments that define and shape them, but the very foundations of all politics?[26]

What we should be clear about is that science—those scientists, researchers and technologists pursuing biotechnological innovations—do not pose and explore these questions as a direct challenge to politics. On the contrary, we may say that for science these questions are epiphenomenal to the extent that they are not necessary for posing or answering those 'scientific' questions required for biotechnological discoveries. Rather, it is the political acceptance, endorsement and incorporation of biotechnological innovation, because it is in some significant way politically 'useful'—in improving health, ameliorating suffering, increasing fertility, expanding food production and therefore fighting poverty—that forces the apparent truth of these scientific insights into the political realm. Once in that realm, however, the political implications of such scientific insights are not at all clear. As we will note below, the extent to which they will transform politics is determined by a range of factors, from the dominance of the 'useful' in modern politics, the specific culture and history of each nation, to the prevalence of the 'industrial-scientific' complex. Nevertheless we can say these scientific insights challenge core aspects of modern politics and importantly, succeed because of the specific means and context of the confrontation. The scientific challenge to politics, at least in biotechnology, is incidental or unintended, a gradual, incremental yet significant redefinition and transformation of politics in scientific terms, fundamentally challenging and potentially undermining the very foundations of all politics. Each new discovery poses novel questions and challenges and in doing so changes and redefines the debate by deploying new language ('genes') and presenting new radical possibilities that shift the terms of the debate. Moreover, introducing scientists as innovators and benefactors alters the grounds for adjudication and resolution of political questions to include the scientific matters for judgment.

## Biotechnology and the limits on political authority

Of course one may argue that these claims of a scientific transformation of political authority are far-fetched or extreme formulations, the stuff of feverish Gothic novels or the new genre of science fiction but nothing more serious. Even if they are not, they certainly do not pose fundamental scientific challenges to politics that cannot be overcome. They seem, indeed, to be ethical questions, difficult and complex to be sure, yet capable of resolution through political means.[27] Politics can address, negotiate and resolve them through regulations and laws and therefore the discussion above has been nothing more than a re-enactment of the 'Frankenstein' and 'Golem' fears that have dogged all stages of modern scientific advance. Still, it would seem that there are major limitations on how politics can meet the challenge of biotechnological innovation. In this section I will explore two such constraints. The first concerns the question of ethics and the moral basis for resisting change; the second concerns the global nature of the political problem, substantially shaped by the role of modern multinational corporations in the development of biotechnological innovations.

### *Politics of theodicy*

Biotechnological challenges to politics are usually interpreted or defined as 'ethical' problems that can be resolved with suitable examination, discussion and debate. The well-known objections to this solution are the obstacles to deliberation due to the lack of public knowledge regarding the scientific advances and their potential political consequences, as well as the limitations of the various attempts to remedy this lack through education campaigns, joint public–expert committees or even deliberative democratic initiatives.[28] But, it is not just the lack of informed judgment and the limits to oversight that constrain political authority. The problem lies in the predominance of the 'useful' in ethical and political debates on technological innovation and change. The history of biotechnology suggests that the 'useful' is the true common ground between politics and science.[29] This common ground at first seems to reveal a weakness in science since it seems to suggest that scientific research that is purely theoretical or abstract will be at a comparative disadvantage to research that has immediate practical or useful applications. Yet in an important sense the primacy of the 'useful' also reveals the weakness of political authority. The useful becomes a powerful means for science to assert and insert its insights into the political domain and in doing so subtly transform it. The reason for this, I would suggest, lies in the moral character of the 'useful', a technological reinterpretation of theodicy that imposes moral demands to accept all useful innovations.

Theodicy, or the question of god's justice, is usually raised when disasters are of such enormity that they seem to challenge divine providence.[30] Modern technological innovation has transformed this question by making the 'useful' a moral obligation on modern political leaders. Two contradictory moral vectors or trajectories appear to characterize modern biotechnology. The first consists of an ever-more comprehensive account of nature that shows in its causality and therefore determinism a world that leaves no room, or denies, human volition, intentionality and therefore moral culpability.[31] This aspect of modern technology calls into question (and arguably continues to undermine) moral, and juridical notions of responsibility, as well as restructuring the political and public policy responses to everyday problems. It transforms, for example, the problem of drug abuse from a moral to a medical question, demanding not censure but cure and rehabilitation (and in some cases accommodation). Such a moral understanding coexists, however, with another contradictory tendency founded upon modern technology that makes us radically responsible for everything.[32] This tendency is based on the simple insight that once we know the cause of something and are able to change it, then the presence or absence of that thing is a matter of our own choice. Greater knowledge makes us, in a sense, participants in what eventuates, and therefore morally accountable for it. This proposition not only reintroduces the question of morality into the world, but radically enlarges its scope. For example, historically it was assumed that there would always be poor people in the world—poverty was a *factum brutum* of being human. All that one could do was to alleviate it—pity, charity, compassion, *noblesse oblige* were the limits of what was possible and therefore morally mandated. But if we can discover the causes of poverty—for example, in inadequate use of farming technology, in predictable weather patterns, in the limitations of badly constructed political institutions—then the continuation of poverty discloses our lack of resolve, our moral inadequacy in solving this problem; it is an indictment of our moral and political will. Thus, the very existence of the problem of poverty, within a specific political community, and indeed anywhere in the world, is proof of our injustice (see, for example, Sen 1995; Drèze *et al.* 1995). If this argument has purchase in the case of poverty, then it will be applicable to almost all those problems that humanity has thought intractable from time immemorial. The practical consequence of this reformulation and resolution of the question of theodicy means that there is a powerful logic, both moral and practical, that insists on the 'useful' innovation as the morally necessary. To deny research into human fertility is to *will* infertility upon childless couples; to reject gene therapy is to *will* blindness, sickness, immobility, and so on. Not allowing science full rein is to will, politically, all of nature's ills that science seeks to cure or overcome. The 'useful' accuses politics of intentional cruelty. In this way modern theodicy has elevated the useful into a powerful trump-card in arguments over biotechnological innovation. In practice it means that the only significant objection to political theodicy has been from religion. But even such resistance is not always consistent. For example, it is true that the strong Catholic presence in Italy and Ireland has meant that these countries have enacted more stringent laws on assisted reproduction technology.[33] Nevertheless, the lack of a clear demarcation between humans and animals in Shinto and Taoism and especially the possibility of reincarnation and transmigration of souls in Hinduism means that researchers in India and China have fewer restrictions on chimera research and xenotransplantation (West 2007: 47, 110). Consequently, theodicy, combined with the ever-present and incremental nature of technical advances, produces a familiar public response to biotechnological innovations: an initial revulsion, a general tendency to object to seemingly unnatural technological changes revealed by the media; followed in time by a general willingness to acknowledge their usefulness; finally resulting in the acceptance of change as uncontentious.[34] This tendency arguably explains the international diversity of political restrictions on biotechnological innovations we noted above. The early development of assisted reproductive technology and its widespread public endorsement has meant that most countries allow industry self-regulation. At the other end of the spectrum we have the almost universal banning of the most recent innovation of reproductive cloning.

### *Industrial-global technology*

The other major obstacle to the political control of scientific innovation lies in the unique relationship between government and business in each nation and how it confronts the multinational nature of biotechnology industry, with the influence of 'Big Pharma', resulting in a 'Science-Industrial complex' (Bud 1993: 189–218; West 2007). Other than outright bans on research, the most common form of national restriction consists of limitation on public funds for experimentation. But the close nexus between government and business in the case of biotechnology means that such control may in some cases be limited, with the potential for industrial influence on governmental policy. The role of the National Institutes of Health (NIH), which assumed responsibility for developing genetic engineering controls in the United States, is instructive. As Wright and Wallace (2000) show, Congressional 'sunshine laws' in the 1960s and 1970s and scientific openness encouraged openness in policy formulation, with public meetings of the Recombinant DNA Advisory Committee (RAC) established to advise the NIH Director on the safety of genetic engineering. But such policy role was soon compromised by the need for industrial secrecy. Part of the problem was that the NIH was not a regulatory agency but a leading sponsor of biomedical research. In addition, private corporations were not bound by NIH controls, which applied only to scientists who received government funding. So while the NIH in 1976 promulgated large-scale culture and release into the environment of genetically engineered organisms as 'prohibited experiments', by the 1980s pharmaceutical industry representatives had modified this requirement to a review of projects by a local bio-safety committee appointed by the company pursuing the project, based on the need to protect trade secrets. Yet even this was challenged by Genentech, which with Eli Lilly proceeded with large-scale production of insulin with genetically engineered microbes without review or approval of RAC, justifying its actions on the need to protect trade secrets. The NIH responded not by seeking Congressional legislative action or reprimanding industry, but by adjusting its politics to conform to industry requirements for secrecy. After experimenting with a 'voluntary compliance scheme' the NIH in the 1980s revised its controls by transferring responsibility for industrial scale uses of engineered organisms to local bio-hazard committees, as advocated by industry representatives. In any case, limiting public funds for experimentation may have the unintended consequence of relinquishing public oversight by forcing controversial research into the private sphere, where there is generally less oversight and control into the nature of research pursued.

The history, culture and religion of a nation inevitably shape its commercial and industrial policy and therefore the political control it will exercise over biotechnology. For example, even though ART is internationally the least regulated form of biotechnology, the legacy of Nazi medical experiments is evident in limits on reproductive therapies imposed by Germany's *Embryo Protection Act 1990* and Austria's 1992 *Act on Procreative Medicine* (West 2007: 38). By contrast, India and China, with substantial populations and extensive poverty making ever-increasing demands on public resources, appear to have fully embraced modern biotechnology, with very few limitations on the nature of research that can be undertaken in each country. India is gaining international recognition due to the limited restrictions it imposes on human clinical trials, where low cost and a ready supply of willing individuals to participate in trials means very few restrictions exist on experimentation. China leads the world on plant biotechnology and, as we saw, the first instance of chimera experiments took place in China. These differences in national biotechnology policy mean that there is wide international divergence in the political control over biotechnology. For example, while China, South Korea, Singapore and Great Britain have the most permissive stem cell policies, Austria, Germany, France and the United States are more restrictive in their funding. Similarly, there is wide diversity in the international treatment of GM foods based on the usefulness of the innovations and fears over the safety of food, with the United States laissez-faire policy on biocrops at one end of the spectrum and European and Japanese strict regulations at the other.[35] Though there are attempts to address national diversity through agreements such as the Cartagena Protocol on Biosafety, such differences give modern multinational biotechnology industry the opportunity to shape domestic policy and where necessary, shift research to new states with more permissive legal regimes for biotechnological research. This form of 'country-shopping' relies on 'scientist-buying', where scientists and sometimes entire laboratories are enticed with lucrative offers to establish and pursue cutting-edge research in new countries with the promise of important commercial advantages. The global nature of modern biotechnology therefore makes it difficult for any one country to limit biotechnological innovation because in doing so it risks losing a potentially lucrative industry to competitors. Embryonic stem cell research is the cornerstone of Singapore's SG$2 billion National Biomedical Science Strategy initiated in 2000. Similarly, the Caribbean island nation of St Kitts has a burgeoning industry in chimera experimentation because monkeys are in abundant supply there, the government imposes few rules on scientists and public and NGO scrutiny is limited (West 2007: 25, 38, 96, 100).

States continue to exercise significant control over biotechnological innovations, but commercial and industrial benefits of biotechnology, the global nature of the scientific and biotechnological industry and the diversity in international policy means that control by any one nation comes at a high price and may in international terms have limited force. The announcement of each new biotechnological discovery cannot help but fuel the persistent and powerful inclination for each state to liberalize its biotechnology laws and policy, and in doing so, effectively allow a scientific incursion into the political.

## Contested authorities

The authority of politics over the science of biotechnology seems obvious in part because it is so visible. Governments have always benefited from biotechnology and governmental support has been crucial for the development of 'new' biotechnology. Governments gain from biotechnological innovations and therefore protect patents and copyright of scientific insights, legislating and regulating what is permitted and demarcating what is not. The history of old and new biotechnology therefore shows the dominance of politics as well as the inextricable link between politics, science and industry. Biotechnology, it would seem, is a handmaiden to politics.

On closer inspection, however, we see a different picture emerge. We need not definitively answer whether science is ultimately a form of politics, or even whether politics can be transformed into a scientific enterprise. What we can say is that biotechnological discoveries are subtly introducing fundamental or constitutive change and innovation into the language of politics, its conceptualisation of core principles and the way it seeks to resolve problems it confronts. The unpredictable, incremental and partial biotechnological insights into the nature of human life, the make-up of the individual and the role of the family challenge the foundations of politics itself. Moreover, scientific authority, if we can use such a political term to describe its influence, does not share the political feature of intentionality and comprehensive vision and purpose. Biotechnology transforms politics incidentally, unintentionally, without strategy or assertion of contradictory authority. It is as though the very presence of scientific insight cannot help but displace or remove political resolutions, assumptions and judgments. Governments appear unable to resist or counter innovations that seemingly dissolve the substance of politics itself. The political focus on the necessary and useful, no longer countered by religious transcendence, seems to provide an irrefutable basis for unconstrained scientific innovation. The global and industrial nature of modern biotechnology in addition exposes modern politics to forces that seem impossible to resist. Consequently, there seems to be no principled way to counter the biotechnological transformation of politics.

The case of biotechnology therefore reveals the increasing power and influence of a scientific way of thinking about the world, not readily visible because we are distracted by the apparently powerful edifice of political authority. But is it specific to biotechnology? Clearly, we would have to pursue the other innovative scientific fields, such as computing, nanotechnology and particle physics to see if biotechnology is unique in this respect. Yet a cursory examination of these new scientific frontiers show familiar patterns, so that even small and limited innovations from all these disciplines would, in their amalgamated and cumulative force, confirm the formidable challenge science in general presents to political authority. The result is a perplexing impression of flux and contradictions, of a cosmos that in its plastic potential increasingly submits itself to our seemingly all-powerful will, while with every new discovery we confirm and testify our essential impotence.

## Notes

1 Lysenko, a fraudulent biologist, preached a *Soviet* science with aspirations of transforming species, and thus ousted geneticists, who were said to preach fatalism and resignation in the face of heredity (Ulam 1973: 643–52). On National Socialist science, see Renneberg and Walker (1994).

2 Of the increasingly radical challenges to the concept of 'science' see, for example, Popper's claim that falsifiability is the essential criterion that demarcates science from non-science, Lakatos's claim that scientific theories are neither verifiable nor falsifiable, Kuhn's argument that science is founded on incommensurable paradigms and Feyerabend's insistence that science is indistinguishable from other ways of knowing. For a general overview of the literature and the debate, see Gauch (2003). Note that

much of this diverse scholarship draws upon the radical critiques by Nietzsche and Heidegger.

3 See Aquinas, *Summa Theologia*, Part 1, Question 1, Reply to Objection 1.

4 See Dewey's defence of pragmatism: Larry A. Hickman, 'Populism and the Cult of the Expert' in Winner (1992: 91–104).

5 This discussion draws primarily on Bucholz (2007, 2010), as well as Hulse (2004), Bud (1993) and Thackray (1998).

6 This finding was anticipated by the discovery of cell theory and sources of fermentation by Theodore Schwann and Cagniard-Latour: see Berche *et al.* (2009). On Pasteur, see his major work Louis Pasteur's *Studies on Fermentation: The Diseases of Beer, Their Causes, and the Means of Preventing Them* (1879).

7 See generally Kohler (1973); Barnett and Lichtenthaler (2001); Hare (1970).

8 On the mass production of penicillin, see Neushul (1993); on the history of penicillin more generally, see Lax (2004); Bud (2007).

9 See generally Bud (1998); Goujon (2001).

10 For an overview, see Bud (1993: 189–210).

11 See the list of major events in Buchholz (2007: 1159).

12 For a history of r-DNA, see Lear (1978); Bucholz and Collins (2011); McRae (2013: 127). For a personal account, see Berg and Mertz (2010).

13 See generally Chandler (2009); Raviña (2011).

14 On the OTA report, see Bucholz (2007: 1157–63).

15 Hall (1987) provides an account of the development of genetically engineered insulin, with competition between groups at Harvard University, the University of California (San Francisco), and the biotech company Genentech.

16 As Wright and Wallace (2000: 47) note, prior to this decision patents could only be obtained for the process that used the microorganism, but not for the microorganism itself, which effectively presumed that life could be patentable.

17 See Huijer (2003). The project is coordinated by the US Department of Energy and the National Institutes of Health, with additional contributions from the Wellcome Trust (UK) as well as France, Germany, China and others. For its main goals, see www.ornl.gov/sci/techresources/Human_Genome/home.shtml (accessed October 4, 2013).

18 For example, one of the first companies to take advantage of the commercial potential of r-DNA was Cetus, started in 1971 by Peter Farley, a physician, Ronald Cape, a biochemist and Donald Glaser, Nobel laureate physicist (Buchholz 2007: 1158).

19 On the politics of technology, see, for example, Scharff and Dusek (2003); Pitt (2000); Kraft and Vig (1988); Street (1992); Feenberg (1999); Katz *et al.* (2003). On the international regulatory framework, see Isasi and Knoppers (2006).

20 See, for example, Buctuanon (2001) regarding globalization and biotechnology; Hall (2009) on the tension between biotechnology and liberal democratic governance, and Grabel and Gruen (2007) on the ethical dimension.

21 For a detailed summary of the nature of these innovations and the contemporary political responses to them see West (2007).

22 See, for example, the discussion in Anderson and Tollefsen (2008). In this context consider the 2015 vote by the UK House of Commons to legalise the use of donor mitochondrial DNA for in-vitro fertilization (IVF) conceptions, allowing the creation of children who will have three biological parents.

23 On the role of major corporations see West (2007: 50). For a critique of commodification due to the influence of modern corporations see generally Lewontin (2000) and Lewontin *et al.* (1985).

24 On the biology of species identity and the morality of crossing species boundaries, see Robert and Baylis (2003).

25 For arguments regarding embryos, eugenics and human nature, see Devolder and Harris (2007); Fenton (2006); Jotterand (2008); Lauritzen (2005); Michael (2001).
26 See, for example, Brodwin (2000), where such questions of genealogy, maternity and ethics are explored.
27 For attempts to engage with these innovations using contemporary concepts such as 'rights', see, for example, Krimsky and Shorett (2005) who advocate a 'genetic bill of rights'.
28 On the politics of inquiries see Peters and Barker (1993); Fischer (2000). On the use of participative or deliberative democracy to overcome these dilemmas, see Forbes (1989); Masters and Kantrowitz (1988); Willard (1996).
29 Consider, for example, the unique and influential Asilomar meeting, a conference that took place in February 1975 in Asilomar State Beach, California. Organized by Paul Berg, the meeting of 140 biologists, lawyers and physicians agreed on guidelines to ensure the safety of r-DNA technology. These voluntary standards adopted the 'precautionary principle' for research, acknowledging practical considerations as an important limit on scientific research (Krismsky 2005: 313; Abels 2005).
30 See, for example, Pierre Bayle's *Various Thoughts on the Occasion of a Comment* (2000).
31 This aspect of modern technology was famously noted by Kant, who sought to overcome the determinism of Newtonian physics, which in positing scientific causality denied human beings their dignity, with a noumenal dignity founded upon autonomous human freedom: see generally *The Metaphysics of Morals* and his political works, such as *On Perpetual Peace*.
32 On the possible religious origin of this moral disposition, see Funkenstein (1986).
33 The Christian insistence that life begins at conception contrasts with the Islamic and Judaic view that ensoulment does not take place until 40 days after fertilisation. On the religious debate in the US, see Green (2010).
34 Maienschein (2001: 431) describes it as the 'cycle of reactions from yuk to tolerance to enthusiasm to reflective concern'. The case of Great Britain is revealing. The initial reluctance with cloning had by 2005 began to abate, allowing British scientists to experiment with therapeutic cloning.
35 See generally Bodiguel and Cardwell (2010); Paarlberg (2013); Pollack and Shaffer (2010).

## References

Abels, G. 2005. 'The Long and Winding Road from Asilomar to Brussels: Science, Politics and the Public in Biotechnology Regulation.' *Science as Culture*, December, 14(4): 339–53.

Anderson, R.T. and Tollefsen, C. 2008. 'Biotech Enhancement and Natural Law.' *The New Atlantis*, spring, 20: 79–103.

Barnett, J.A. and Lichtenthaler, F.W. 2001. 'A History of Research on Yeasts 3: Emil Fischer, Eduard Buchner and Their Contemporaries, 1880–1900.' *Yeast*, March, 18(4): 363–88.

Bayle, P. 2000. *Various Thoughts on the Occasion of a Comment*, trans. R. Bartlett. New York: SUNY Press.

Berche, B., Henkel, M. and Kenna, R. 2009. 'Critical Phenomena: 150 Years Since Cagniard de la Tour.' *Journal of Physical Studies*, 13(3): 3001.

Berg, P. and Mertz, J.E. 2010. 'Personal Reflections on the Origins and Emergence of Recombinant DNA Technology.' *Genetics*, January, 184(1): 9–17.

Bodiguel, L. and Cardwell, M. (eds). 2010. *The Regulation of Genetically Modified Organisms: Comparative Approaches*. Oxford: Oxford University Press.

Brodwin, P.E. (ed.) 2000. *Biotechnology and Culture: Bodies, Anxieties, Ethics*. Bloomington, IN: Indiana University Press.

Buchholz, K. 2007. 'Science—or Not? The Status and Dynamics of Biotechnology.' *Biotechnology Journal*, 2: 1154–68.

Buchholz, K. 2010. 'History of Biotechnology.' In *Encyclopedia of Industrial Biotechnology: Bioprocess, Bioseparation, and Cell Technology*. M.C. Flickinger (ed.). Milton: Wiley, 2856–70.

Buchholz, K. and Collins, J. 2011. *Concepts in Biotechnology: History, Science and Business*. Weinheim: Wiley-VCH.

Buctuanon, E.M. 2001. 'Globalization of Biotechnology: The Agglomeration of Dispersed Knowledge and Information and its Implications for the Political Economy of Technology in Developing Countries.' *New Genetics and Society*, April, 20(1): 25–41.

Bud, R. 1993. *The Uses of Life: A History of Biotechnology*. Cambridge: Cambridge University Press.

Bud, R. 1998. 'Molecular Biology and the Long-Term History of Biotechnology.' In *Private Science: Biotechnology and the Rise of the Molecular Sciences*. Arnold Thackray (ed.). Philadelphia, PA: University of Pennsylvania Press, 3–19.

Bud, R. 2007. *Penicillin: Triumph and Tragedy*. Oxford: Oxford University Press.

Capps, B.J. 2008. 'Authoritative Regulation and the Stem Cell Debate.' *Bioethics*, January, 22(1): 43–55.

Capps, B.J. and Campbell, A.V. (eds). 2010. *Contested Cells: Global Perspectives on the Stem Cell Debate*. London: Imperial College Press.

Caulfield, T. 2003. 'Human Cloning Laws, Human Dignity and the Poverty of the Policy Making Dialogue.' *BMC Medical Ethics*, 4: 3–9.

Chandler, A.D. 2009. *Shaping the Industrial Century: The Remarkable Story of the Evolution of the Modern Chemical and Pharmaceutical Industries*. Cambridge, MA: Harvard University Press.

Devolder, K. and Harris, J. 2007. 'The Ambiguity of the Embryo: Ethical Inconsistency in the Human Embryonic Stem Cell Debate.' *Metaphilosophy*, April, 38(2–3): 153–69.

Drèze, J., Sen, A. and Hussain, A. (eds). 1995. *The Political Economy of Hunger: Selected Essays*. Alderley: Clarendon Press.

Feenberg, A. 1999. *Questioning Technology*. London: Routledge.

Fenton, E. 2006. 'Liberal Eugenics and Human Nature: Against Habermas.' *Hastings Center Report*, November/December, 36(6): 35–42.

Feyerabend, P. 1975. *Against Method*, 3rd edn. London: Verso.

Fischer, F. 2000. *Citizens, Experts and the Environment: The Politics of Local Knowledge*. Durham, NC: Duke University Press.

Forbes, H.D. 1989. 'Dahl, Democracy, and Technology.' In *Democratic Theory and Technological Society*. R. Day, R. Beiner and J. Masciulli (eds). London: M.E. Sharpe, 278–305.

Funkenstein, A. 1986. *Theology and the Scientific Imagination from the Middle Ages to the Seventeenth Century*. Princeton, NJ: Princeton University Press.

Gauch Jr, H. 2003. *Scientific Method in Practice*. Cambridge: Cambridge University Press.

Goujon, P. 2001. *From Biotechnology to Genomes: The Meaning of the Double Helix*. Singapore: World Scientific Publishing.

Grabel, L. and Gruen, L. 2007. 'Introduction: Ethics and Stem Cell Research.' *Metaphilosophy*, April, 38(2–3): 137–52.

Green, R.M. 2010. 'Political Interventions in U.S. Human Embryo Research: An Ethical Assessment.' *The Journal of Law, Medicine & Ethics*, summer, 38(2): 220–8.

Hall, L.K. 2009. 'Biotechnology and the Problem of Liberal Democratic Governance.' *Perspectives on Political Science*, 38(3): 167–72.
Hall, S.S. 1987. *Invisible Frontiers: The Race to Synthesize a Human Gene*. Oxford: Oxford University Press.
Hare, R. 1970. *The Birth of Penicillin and the Disarming of Microbes*. London: George Allen & Unwin.
Hickman, L.A. 1992. 'Populism and the Cult of the Expert.' In *Democracy in a Technological Society*. L. Winner (ed.). Netherlands: Kluwer Academic Publishers, 91–104.
Huijer, M. 2003. 'Reconsidering Democracy: History of the Human Genome Project.' *Science Communication*, June, 24(4): 479–502.
Hulse, J.H. 2004. 'Biotechnologies: Past History, Present State and Future Prospects.' *Trends in Food Science & Technology*, January, 15(1): 3–18.
Isasi, R.M. and Knoppers, B.M. 2006. 'Mind the Gap: Policy Approaches to Embryonic Stem Cell and Cloning Research in 50 Countries.' *European Journal of Health Law*, 13(1): 9–25.
Jotterand, F. 2008. 'Beyond Therapy and Enhancement: The Alteration of Human Nature.' *NanoEthics*, April, 2(1):15–23
Katz, E., Light, A. and Thompson, W. (eds). 2003. *Controlling Technology: Contemporary Issues*, 2nd edn. Amherst: Prometheus Books.
Kohler Jr, R.E. 1973. 'The Enzyme Theory and the Origin of Biochemistry.' *Isis*, June, 64(2): 181–96.
Kraft, M. and Vig, N. 1988. *Technology and Politics*. Durham, NC: Duke University Press.
Krimsky, S. 2005. 'From Asilomar to Industrial Biotechnology: Risks, Reductionism and Regulation.' *Science as Culture*, December, 14(4): 309–23.
Krimsky, S. and Shorett, P. (eds). 2005. *Rights and Liberties in a Biotech Age: Why We Need a Genetic Bill of Rights*. Lanham, MD: Rowman & Littlefield.
Lauritzen, P. 2005. 'Stem Cells, Biotechnology, and Human Rights: Implications for a Posthuman Future.' *Hastings Center Report*, March–April, 35(2): 25–33.
Lax, E. 2004. *The Mold in Dr. Florey's Coat: The Story of the Penicillin Miracle*. New York: Henry Holt & Company.
Lear, J. 1978. *Recombinant DNA: The Untold Story*. New York: Crown Publishing.
Lewontin, R.C. 2000. *The Triple Helix: Gene, Organism, and Environment*. Cambridge, MA: Harvard University Press.
Lewontin, R.C., Rose, S. and Kamin, L.J. 1985. *Not in our Genes: Biology, Ideology, and Human Nature*. New York: Pantheon Books.
Maienschein, J. 2001. 'On Cloning: Advocating History of Biology in the Public Interest.' *Journal of the History of Biology*, winter, 34(3): 423–32.
Masters, R. and Kantrowitz, A. 1988. 'Scientific Adversary Procedures: The SDI Experiments at Dartmouth.' In *Technology and Politics*. M. Kraft and N. Vig (eds). Durham, NC: Duke University Press, 278–305.
McRae, W.P. 2013. *The Visioneers: How a Group of Elite Scientists Pursued Space Colonies, Nanotechnologies, and a Limitless Future*. Princeton, NJ: Princeton University Press.
Michael, M. 2001. 'Technoscientific Bespoking: Animals, Publics and the New Genetics.' *New Genetics and Society*, 20(3): 205–24.
Montpetit, É., Rothmayr, C. and Varone, F. (eds). 2007. *The Politics of Biotechnology in North America and Europe*. Lanham, MD: Lexington Books.
Neushul, P. 1993. 'Science, Government and the Mass Production of Penicillin.' *Journal of the History of Medicine and Allied Sciences*, 48(4): 371–95.

Paarlberg, R. 2013. *Food Politics: What Everyone Needs to Know*. Oxford: Oxford University Press.

Pasteur, L. 2014 [1879]. *Louis Pasteur's Studies on Fermentation: The Diseases of Beer, Their Causes, and the Means of Preventing Them*. Whitefish, MT: Kessinger Publishing.

Peters, B.G. and Barker, A. (eds). 1993. *Advising West European Governments: Inquiries, Expertise and Public Policy*. Pittsburgh, PA: University of Pittsburgh Press.

Pitt, J.C. 2000. *Thinking about Technology: Foundations of the Philosophy of Technology*. New York: Seven Bridges Press.

Pollack, M.A. and Shaffer, G.C. 2010. *When Cooperation Fails: The International Law and Politics of Genetically Modified Foods*. Oxford: Oxford University Press.

Raviña, E. 2011. *The Evolution of Drug Discovery: From Traditional Medicines to Modern Drugs*. Weinheim: John Wiley & Sons.

Renneberg, M. and Walker, M. (eds). 1994. *Science, Technology, and National Socialism*. Cambridge: Cambridge University Press.

Robert, J.S. and Baylis, F. 2003. 'Crossing Species Boundaries.' *The American Journal of Bioethics*, summer, 3(3): 1–13.

Scharff, R.F. and Dusek, V. (eds). 2003. *Philosophy of Technology: The Technological Condition—An Anthology*. Malden: Blackwell Publishers.

Sen, A. 1995. *Inequality Reexamined*. Oxford: Clarendon Press.

Street, J. 1992. *Politics and Technology*. New York: Guilford Press.

Thackray, A. (ed.) 1998. *Private Science: Biotechnology and the Rise of Molecular Sciences*. Philadelphia, PA: University of Pennsylvania Press.

Ulam, B.A. 1973. *Stalin: The Man and His Era*. London: Allen Lane.

West, D.M. 2007. *Biotechnology Policy Across National Boundaries: The Science-Industrial Complex*. Houndmills: Palgrave Macmillan.

Willard, C.A. 1996. *Liberalism and the Problem of Knowledge: A New Rhetoric for Modern Democracy*. Chicago: University of Chicago Press.

Wright, S. and Wallace, D.A. 2000. 'Varieties of Secrets and Secret Varieties: The Case of Biotechnology.' *Politics and the Life Sciences*, March, 19(1): 45–57.

# 10 Conclusion

## A democratic tension?

*Michael Heazle and John Kane*

In the Introduction to this volume, we framed our contribution to the existing literature in the context of three waves of scholarship (Collins and Evans 2002) that addressed the nature and interaction of expert and political authorities, including the normative question of how they *should* interact given liberal democratic requirements for 'legitimate' policy. The linear, decisionist policy model of early wave one accounts, which assumed that expert authority derived directly from objective scientific knowledge and operated more or less passively in the service of political decision-makers, was rejected by wave two researchers who argued that the production and application of scientific knowledge could not be entirely separated from the influence of political values and, in some instances, policy bias. In problematizing the wave one account, wave two argued the inherently political nature of expertise, calling into question its claims to authority based on privileged access to objective knowledge. The effect (if not precisely the intention) was to collapse political and expert forms of authority into one another. Wave three responded by attempting to preserve the original distinction as somehow necessary while addressing legitimate issues of political bias raised by wave two.

Certainly, wave two conclusions made it difficult to explain, or even identify, instances where the two forms of authority nevertheless seemed to remain distinct and to operate in ways largely consistent with wave one expectations. Such instances may, in fact, be common in relatively low stakes, day-to-day policy making but also may be found in situations like post-tsunami Japan where decision-makers faced extremely high stakes in attempting to balance both physical security and socioeconomic security concerns (as Paul Scalise's chapter in this volume shows). This suggests that attempts within wave three to maintain the validity of a real distinction between scientific and political authority is well-founded. Indeed, it is ironic, as Darrin Durant observes in his chapter, that the main prescriptive force of wave two's analysis has been channeled into a demand for more transparent negotiation of the expertise-politics boundary to make it more nearly resemble, if for very different reasons, the ideal separation of science and policy figured in wave one.

The arguments made by wave two Science and Technology Studies (STS) challenging science's traditional claims to objectivity and autonomy provide important and well-founded insights into both the nature and limits of science and the knowledge it produces. The more extreme constructivist positions, however, seem a step too far insofar as they tend to reduce expert authority to just a variant form of politics. The sciences have established means of moderating disputes within their fields (peer review, academic exchange and debate, falsification trials and so on) allowing them to adjudicate progressively on all knowledge claims formally presented. It is these disciplinary standards that alone give validity and respectability to scientific endeavor and, moreover, assure it of a welcome in the policy process. These standards may never themselves be beyond the possibility of challenge (see, for example, Kuhn 1996), but, as some STS scholars also have argued, the major challenge confronting the supposed apolitical authority of expertise occurs not in the laboratory as such but in the highly politicized domain of policy making.

Indeed, when expert claims enter the policy realm, particularly those that are not backed by expert consensus, they are vulnerable to becoming 'politicized' by being tied to positions supporting or opposing one or another policy option. Yet to declare the authority of experts suspect or unreliable for this reason alone is setting the bar too high given the necessarily political nature of the policy process, and the inability of experts to control the political ends to which their research or advice may be employed by others. In an ideal world (the world of wave one), political authorities will consider seriously what experts say is the case if they wish to make sensible policy decisions that have a chance of working. This ideal often goes unrealized, which is, of course, what all the debate of the subsequent waves is about. And there are undoubtedly issues of such inherent uncertainty and complexity that no expert consensus exists, yet which are so urgent that the policy process demands a decision. In such cases experts have no choice but to resolve disagreement through negotiation over what is more or less likely, an act of professional judgment rather than an assertion of positive knowledge.

Nevertheless, science, we may say, generally argues about what *is* the case according to available evidence; politics, meanwhile, argues about what *should* be the case according to political will, however that may be determined. This elemental distinction must be retained—despite the now well-documented overlaps that have occurred between them in policy making—if science is not to be falsely reduced to politics, on one hand, or politics to science, on the other. Neither scenario is anyway compatible with what liberal democracies demand for policy to be legitimate.

So although the legitimacy of expert authority in policy rests on the presumed non-political nature (in the sense of unsullied by partisan political disputes) of expertise and knowledge, such claims to objectivity do not need to imply that expertise operates in a political vacuum—wave two research has demonstrated this is clearly not the case—or that expert claims are uncontestable, since permanent openness to contestability is one of the hallmarks of science. What ultimately distinguishes science-based expert authority from political authority is neither its uncontestability, nor its detachment from values, but rather that knowledge disputes *among* experts—as opposed to disputes *over* expert knowledge in policy debate—are resolved (if they are) not primarily through bargaining, compromise or compulsion as in politics, but ultimately by appeal to the institutionalized standards and criteria that guide disciplinary inquiry.

If it can be shown, for a particular discipline, that judgments and findings are not primarily driven by accepted standards then its entire status as a genuine source of policy relevant expertise is called into question. This is part of the argument presented in this volume by John Kane with regard to neoclassical economics, which naively identifies science with modeling per se, paying scant regard to the plausibility or empirical reality of the assumptions its models employ. The authority provided by apparently rigorous models has nevertheless enabled a significant influence on public policy because prescriptions are claimed to be a result of hard scientific analyses that allegedly admit of no (rational) alternative. At the heart of the neoclassical agenda, however, is a simplistic conception of 'naturally' free markets that decidedly loads the dice in favor of particular policies.

Kane's chapter suggests, however, that the interaction of political and expert authorities in the case of economics is exceedingly intimate and that their fortunes tend to rise and fall together. Indeed, any overly sharp separation between economics (of whatever stripe) and politics may be impossible, since economic issues automatically bear on political ones, and economic values on political values, particularly given the modern centrality of 'economic management' for political leaders. This is not to say that genuine economic knowledge is impossible, but only that it may be inextricably bound up with forms of political knowledge. For example, the question, 'Is it possible simultaneously to maintain high employment, stable prices, and rapid growth?' is surely at once an economic and a political question, yet one that nevertheless admits of reasoned argument in which expertise is highly relevant. This may be to say that economics is always in reality *political* economics, and that a problem exists only if political preferences are concealed within an apparently objective 'scientific' apparatus. Although we tend to assume that expert opinion is most politically potent, ironically, when it is perceived as *not* politically motivated, it is hard, and maybe impossible, to think of an economist of any school who does not hold definite and usually strong political positions which generally set the parameters of their economic analyses.

Governments will undoubtedly continue to trade on a sharp distinction between science and politics, having little choice but to appeal to the evidence of supposedly neutral experts to support their policies.[1] But whether economics is a special case or whether close interaction is characteristic of all the social sciences to a greater or lesser degree, given their well-known commitment to human and social values, is an interesting question. At any rate, relations between expert and political sources of authority will inevitably be complex, particularly where policy issues have become sites of sustained political contest. Wave three, broadly speaking, attempts to manage this complexity by taking a 'middle way' between waves one and two. It recognizes that, whatever the complications afflicting the wave one description of the policy process, the separation of expert authority from political authority remains indispensable to policy making in a modern, technology-dependent society. The central issue for wave three scholars has thus been the question of how expert authority, given its inability to remain entirely apolitical and neutral in the policy field, can legitimately influence public policy while at the same time ensuring the primacy of democratic values and representation. Their response generally has been to advocate better oversight of some recognized form of expert influence through broader participation in the determination of the kinds of evidence and expertise to be used in policy making.

This sounds theoretically attractive, but how does it fare in the light of the assembled chapters? In some policy areas the prospect of resolving expert-political tensions through greater public participation may be real. A rather extreme case is provided by Daniel Sarewitz and David Kriebel in their account of how science's inability to guide the US government on effective chemical regulation led to regulatory cooperation between consumers and corporations, including global giants such as Walmart and Staples. This new model represents a surprising amalgam of political strategies, bringing together the concerns of the environmental health and safety world with the corporate world's preference to act voluntarily rather than via regulatory mandates. Whatever the problems with this emerging approach, it has the enormous virtue of actually leading to action in ways that are democratically responsive, and politically and economically plausible. It provides an instance of 'public' regulation that resolved the tensions that had emerged between expert and public perceptions of risk by sidestepping government regulation efforts and replacing them with democratically driven arrangements between civil society and corporations.

In other cases, however, public oversight of expertise would seem difficult or impossible to implement, as well as beside the point. Paul Pillar's analysis of how policy elites in the US interpreted and used secret intelligence to justify military action in the 2003 Iraq War indicates that critical areas of public policy are likely to prove highly resistant to moderation through more inclusive public oversight. And, note, the issue here was not the malign influence of expert testimony over democratic representatives, but the use (or misuse) of expert intelligence precisely *by* democratic leaders. The episode showed the dominance of political over expert authority in decisions to use armed force, which, Pillar argues, flows largely from the particular characteristics of national security issues, especially the limited public release of classified information and the politically charged nature of any decision to go to war. Intelligence experts were simply incapable of shaping or reordering the political priorities of a headstrong executive. The makers of the Iraq War were motivated largely by a desire to use regime change in Iraq to spread democratic, free market values throughout the Middle East and were, in Pillar's view as a senior US security advisor, operating according to 'gut feelings' about these values. They were not about to be stopped by inconvenient facts and assessments about near-term prospects for Iraq.

It is instructive to recall that this administration rejected the whole idea of expert evidence and advice as a small-minded obsession of what it contemptuously referred to as the 'reality-based community', believing that the use of American power would create its own reality on the ground. This turned out to be true, but not in the positive sense the administration had imagined. But however contemptuous some in the Bush administration were of 'the reality crowd' (Suskind 2004), it nevertheless continued to try to establish the case for war on the basis of expert evidence and analysis, thereby demonstrating the importance of expert authority as a source of policy legitimacy. The whole episode, however, stands as an object lesson on the importance of heeding expert testimony if one wishes to make sound policy and avoid egregious political folly. The point is that broader public or perhaps Congressional oversight of expert evidence would not have prevented disaster in this case as the responsibility for it lay wholly with the political masters and the ways in which they chose to interpret the variety of intelligence advice that was on offer.

An entirely different dynamic is evident in Paul Scalise's examination of the roles played by expert and political authority in the decisions to restart nuclear reactors in some regions of Japan but not in others. Of the three proposed reactor restarts, and despite the high stakes nature of the decisions and the many uncertainties involved, only one of the cases was characterized by any significant tension between independent expert panels and political decision-makers. In two of the cases, tensions failed to emerge due to a lack of opposition either from the public, industry or government to the expert advice on offer. In one case that advice was to restart (Hokkaido), the other to not restart (Tsuruga), and the final decisions were consistent with these recommendations. Importantly, there was no scientific disagreement among the experts in either of these cases. Despite the fact that there were clear incentives in the non-restart case for the nuclear industry to attempt to undermine the expert consensus, this did not occur (a strike against the idea of wholesale regulatory capture by the industry). In the third instance, contrastingly, no expert consensus existed, and it appears that tensions were successfully managed by the political executive through negotiation and assessment of the public interest (i.e. risk of another nuclear accident versus energy security and socioeconomic risk).

All three decisions could be described as being consistent with the wave one model of expert influence in policy making, with decisions made in the public interest by political authorities drawing on the independent advice of experts who are 'on tap and not on top'. An interesting question is why that model appeared to hold in this case, and not be muddled by the machinations of special interests or headstrong politicians. One possible explanation is that in situations where the stakes are potentially 'life or death', a fact underlined by recent Japanese experience of the reality of the risks involved, no one was willing to try to undermine expert authority by 'playing politics'.[2]

In another case presented here, Michael Heazle's discussion of the International Whaling Commission (IWC), the policy relevance and legitimacy of expert knowledge claims appeared to be judged (by both policy actors and experts) in terms of their fit with particular political values rather than their reliability or scientific validity. This was not, as in the US intelligence case, only a matter of policy makers 'cherry-picking' evidence to suit pre-existing priorities but also an example of how the influence of expert authority can wax and wane over time as broader political priorities and circumstances change. The IWC was the first international wildlife management regime to explicitly make science the exclusive basis for policy making, but that science was predicated on the survival and revival of whale stocks after over-hunting, with the long-term aim of continuing whaling on a sustainable basis. In the intervening years, however, the international debate shifted from sustainability to the morality of whaling *tout court*, in which circumstances the Scientific Committee's expert authority became largely irrelevant to policy marked by sustained levels of intense political conflict. The IWC remains institutionally unable to resolve its now-confused role, with sharp divisions opening up among both scientists and policy actors over the kinds of science that should legitimately inform that role. Heazle argues that the IWC Scientific Committee's relevance to policy has been almost entirely determined by the political values its recommendations and findings are seen either to support or conflict with. Far from realizing its original role under the IWC convention of 'speaking truth to power', scientific expertise in the commission has for the most part been relegated not only to being 'on tap' but also to being 'in line', thereby creating an ongoing crisis of policy legitimacy in the commission.

The IWC case is ultimately one in which moral values (saving the whale) render scientific judgments (about the viability of whale stocks) largely irrelevant. In similar but also different fashion, Clare Heyward and Steve Rayner's assessment of how proposed geoengineering solutions to climate change have been perceived and debated illustrates how fundamental democratic values can overawe expert policy advice, regardless of scientific rigor, if the latter is seen to threaten those values. The debate has pitted social scientists (among others) against physical scientists who argue the case for geoengineering. A major criticism relates to the suggested use of stratospheric sulphate aerosols (SSA) to block sunlight. On the presumption that this process will need to be governed globally, social scientists have raised questions concerning the democratic legitimacy of such governance. However, as this chapter points out, there already exist global institutions addressing climate change whose legitimacy has rarely if ever been questioned by social scientists in mainstream climate discourse. This asymmetry raises a question about the role and methodology of the social scientists examining public perceptions of geoengineering technologies. Heyward and Rayner suggest that their critical attitude towards geoengineering is based on a defense of democratic values in the face of 'technological essentialism', the idea that what are essentially political problems can be solved with a technological fix. Their case indeed illustrates what is most at stake in the arguments of all three 'waves' of thought on expert-political interaction, namely who in the end gets, or should get, the determinative say in important policy matters (democrats or technocrats).

Overall, these cases portray the complexity and variability of the interactions between political and expert authority, though with a tendency to demonstrate the dominance of the political over the sphere of expertise. And if political authority indeed normally dominates, it is probably because political preferences reflecting either mainstream values, or the priorities of the policy elite, or of powerful special interests, have preset the parameters of what can count as legitimate policy knowledge. Even in Kane's economic case, politicians are not passive dupes of economists' theories but willing accomplices won over, prior to the Global Financial Crisis, by promises of risk-free finance that fueled apparent, but ultimately false, prosperity (in the form of home ownership in conditions of rising house values) for their democratic constituents. And if he is correct in assessing contemporary economic 'science' as really a disguised politics, then the political wins again.

Second-wave theorists, we have indicated, tended to discover disguised politics in all forms of expertise, but this seems too sweeping an assumption.[3] *Pace* those theorists, the important question in understanding the nature and occurrence of expert–political authority tensions is not whether all expertise is inherently political, but rather when and how expertise becomes a site of political contest and, as a consequence, *politicized.* It may be true that achieving a negotiated demarcation between the two forms of authority is not possible when they are rendered indistinguishable, either through the politicization of expert claims or the 'scientization' of political questions and debate. Yet it is surely in the nature of political contest, conflict and debate that expert opinions more readily become politicized than politics ever becomes scientized—except, of course, in those cases where political protagonists use disputes among experts to obscure the values actually at stake. The primacy of the political may be more difficult to achieve if some level of expert consensus exists, but in fact perfect consensus is seldom if ever found. Thus, in contrast to Habermasian fears of the technocratic hijack of political authority through the scientization of public policy as societies become more technologically advanced and dependent, it is likely that expert authority will more often be hijacked by political interests that selectively support one expert position over another on the basis of its ability to *fit with* policy rather than its ability *to inform* policy.

It would seem then to be an impossible task to eliminate the possibility of political–expert tensions emerging, for there is always the potential for disagreement between what experts say must be the case and what politicians claim should be (or determine will be) the case. We have noted that legitimate policy making in a liberal democracy implies policy broadly recognized to be informed by both expert and political forms of authority. It need hardly be said that the congruence of these two forms of legitimacy can never be automatically assumed, and our cases show the processes of combining them can be more or less well managed. But managed they must be. Darrin Durant's chapter examines the collapse of bee colonies to interrogate the theories of the three waves, and comes to the tentative suggestion that there may be value, and democratic value, in preserving the tension between political and expert authority instead of trying to resolve it. We would concur and state positively that such tensions are not something to be eliminated but an inevitable aspect of the democratic political process. Good policy making, if and when it occurs, cannot result from resolving the tension through either political suppression of expertise or technocratic domination, but rather through contest, conflict, debate and negotiation among the parties. Politics being what they are, this process is inevitably messy, tardy and often inconclusive but there exists no ideal alternative in a non-authoritarian political system. What is certain is that democratic doctrine demands that political authority must reign supreme in the end, even if that leads to the judgment that 'bad' politics has trumped 'good' policy.

Let us conclude, however, with the wild card in our pack of cases, that of Haig Patapan and the issue of biotechnology. In Patapan's analysis of matters like reproductive technology, genetically modified foods, stem cell research and cloning, technology promises (or threatens) to emerge eventually triumphant over politics, if in a somewhat sideways manner. Though politics may significantly shape and influence biotechnological research, scientific innovations in turn transform political life by, for example, altering the meaning of family relations and even ultimately our conception of what it means to be 'human'. These changes may be incremental but they are far-reaching, and can radically undermine political authority because they are difficult to resist politically. This is partly because of the transnational character of biotech industry and science, but mostly (and here is the democratic element) because of the perceived benefits of the innovations to many ordinary people. The authority granted to science here, in other words, comes from the desirable things it brings to people's lives rather than from perceptions of its claim to objectivity. The changes take place incidentally, without specific purpose or strategic ambition, but the long-term unintended result is that technologies capable of reshaping our identities, values and interests proceed in a manner beyond the comprehensive control of conventional political authority. This is a radical thought indeed (that perhaps echoes the observations of Hajer (2003)), and one that casts our central question of the tension between political and expert authority in a quite different and disturbing light.

Patapan's provocative intervention notwithstanding, our overall conclusion is that the potential for tension between the two forms of authority always exists, and that it is a mistake to try to prevent its emergence during disagreements over what is and what should be. Such disagreements can, and indeed must be, resolved in liberal democracies through the negotiation of a broadly acceptable balance between both forms of authority if the goal of legitimate (read 'good') public policy is to be realized. Among our cases, as we saw, such agreement occurred, for example, in Japan's nuclear restarts and also the public regulation of chemicals in the US.[4] In others, however—the 2003 invasion of Iraq, the IWC, economic policy and the geoengineering debate—the two forms of authority were blurred, making the negotiation of a balance much more difficult and possibly setting the stage for crises of policy legitimacy.

The bottom line appears to be, however, that a balance cannot be negotiated if the two forms of authority are not kept in some sense distinct. After all, even when political authority is delegated to experts, their status as experts remains distinct from the authority that empowers them to determine policy. The source of, and capacity to claim, expert authority will always reside in learning and practice in some disciplinary field; the source of political authority always lies quite elsewhere. Our cases indicate that tension occurs when the values shaping what *should or will* be the case clash with knowledge/professional judgements attempting to guide thinking on what is/can be the case. It may be the expectation in technocratic Singapore that, since experts have political authority, few such tensions will occur. In liberal democracies, no such unanimity can be automatically expected, especially when stakes are high under conditions of considerable uncertainty. In such cases our findings indicate that the politicization of expert input and/or the overextension by experts of their authority into politics are likely to cause serious entanglement and blurring of categories. Nevertheless, no clarity will be gained by abandoning broadly accepted demarcations, or from ceasing to try to sort out who says and does what and with what authority.

Though our collection of cases has been necessarily limited, we hope that our studies may provide fruitful avenues for further research into how the value of expertise, essential to liberal democratic political systems, may be better and more consistently realized.

## Notes

1 Indeed, the continued currency of positivist-based notions of expert authority and the linear-rationalist process through which it is expected to help inform and guide policy is demonstrated by ongoing pledges from political leaders to provide 'evidence-based policy' and ongoing endorsements, although usually selectively applied, of science 'speaking truth to power,' as demonstrated, for example, by United Nation's Secretary General Ban Ki-Moon's 2014 assertion (*The Economist*, November 3, 2014) that international action on climate change can no longer be delayed now that 'the science has spoken'.

2 Part of the answer may also lie with the nature of Japan's political culture, which manifests habitual deference to expert authority when important and urgent decisions need to be made.

3 Most disciplines can show some sign of progressive knowledge over time, sometimes with revolutionary turns, which evidences some purchase on reality, but economics may be a special case. An Alfred Marshall reborn after a century might not find much of real substance in current neoclassical thought to add to his foundational *Principles of Economics*.

4 We suspect that myriad cases of day-to-day policy making, which go largely unnoticed, may contain instances where a balance between political imperatives and expert advice was achieved. Indeed, examining cases where there is no ongoing crisis of legitimacy may be one useful direction for further research into the conditions under which tensions may be successfully negotiated.

## References

Collins, H.M. and Evans, R. 2002. 'The Third Wave of Science Studies: Studies of Expertise and Experience.' *Social Studies of Science*, 32(2): 235–96.

Hajer, M. 2003. 'Policy Without Polity? Policy Analysis and the Institutional Void.' *Policy Science*, 36(2): 175–95.

Kuhn, T. 1996. *The Structure of Scientific Revolutions*, 3rd edn. Chicago: University of Chicago Press.

Suskind, R. 2004. 'Faith, Certainty and the Presidency of George W. Bush.' *The New York Times Magazine*, October 17. Retrieved at: www.nytimes.com/2004/10/17/magazine/17BUSH.html?_r=0.

# Index